Ren Jishun (Jen Chi-shun) Jiang Chunfa
Zhang Zhengkun   Qin Deyu

Under the Direction of Professor Huang Jiqing (T.K.Huang)

Geotectonic Evolution of China

With 30 Figures, 14 Plates

Science Press Beijing
Springer-Verlag
Berlin Heidelberg New York
London Paris Tokyo

Ren Jishun (Jen Chi-shun) Jiang Chunfa Zhang Zhengkun Qin Deyu
Under the direction of Professor Huang Jiqing (T. K. Huang)

Institute of Geology, Chinese Academy of Geological Sciences
Beijing, China

Responsible Editor   Li Qifang

Published by Science Press Beijing

Distribution rights throughout the world, excluding The People's Republic of China, granted to Springer-Verlag Berlin Heidelberg New York London Paris Tokyo

Additional material to this book can be downloaded from http://extras.springer.com.

ISBN 978-3-642-64874-8   ISBN 978-3-642-61574-0 (eBook)
DOI 10.1007/978-3-642-61574-0

This work is subject to copyright. All rights are reserved, whether the whole or part of the material is concerned, specifically those of translation, reprinting, re-use of illustrations, broadcasting, reproduction by photocopying machine or similar means, and storage in data banks. Under § 54 of the German Copyright Law where copies are made for other than private use, a fee is payable to "Verwertungsgesellschaft Wort", Munich.

© Science Press Beijing and Springer-Verlag Berlin Heidelberg 1987
Softcover reprint of the hardcover 1st Edition 1987

2 1 3 2 / 3 1 4 0 -5 4 3 2 1 0

Science Press Book No. 4 8 4 5 -8 5

FOREWORD

Anyone studying the geology and tectonics of China and who is not able to read Chinese will need to have a copy of this book: *Ceotectonic Evolution of China* and a companion copy of the *Tectonic Map of China*, scale 1:4000,000. Professor Huang Jiqing and his collaborators from the Institute of Geology, Chinese Academy of Geological Sciences have provided the English-speaking earth scientists with an extremely valuable tool that can be used towards understanding the geology of China. The introductory chapter is necessary to read prior to effective use of the material discussed in other chapters as it clearly presents the philosophy of this school of tectonics. The collaborators of the book acknowledge the important changes have been brough about by the plate tectonics theory but do not fully incorporate these ideas into their discussion.

The book and map are testimony to the tremendous amount of geologic work accomplished by Chinese geologists in the past fourty years. As our Chinese colleagues begin to publish more English summaries such as this, it will become apparent to the rest of the world the vast amount of geologic mapping along with supportive stratigraphy and geophysics that has already been accomplished. Nearly all major non-Chinese tracts on tectonic synthesis of the world treat China in only a cursory fashion because so little is known of the area. With this text and map, future world tectonic synthesis can no longer afford to leave China out of the picture.

One of the fascinating aspects of Earth Science in China today is the philosophy stimulated by Chairman Mao who encouraged "a hundred schools of thought to contend". Professor Huang Jiqing provides here a viable exposition on tectonic cycles, geosynclines, and deep fractures as illustrated by the evolution of Pal-Asia from Precambrian to Paleozoic. Beginning with the Indosinian orogenic cycle (Triassic), Professor Huang then introduces plate movements to explain the marginal-Pacific and Tethys-Himalayan tectonic domains evolution up to the present time. In contrast to these publications, Professor Li Chunyu and his colleagues also from the Institute of Geology, Chinese Academy of Geological Sciences explain the *Tectonic Map of Asia,* 1982 (scale, 1:8,000,000) in terms of plate tectonics. The fact that these separate tectonic schools of thought coexist in the same institute and use the same basic geologic data is a good illustration that geologic research can profit from allowing multiple working hypothesis to obtain.

Robert G. Coleman
Professor of Geology
Stanford University
Stanford, California, USA
Dec. 30, 1986.

PREFACE

Since the publication of the *Geology of China* by Li Siguang (J. S. Lee) in 1939 and *On Major Tectonic Forms of China* by Huang Jiqing (T. K. Huang) in 1945, great progress has been made in the study of geotectonics in China. Particularly since the founding of new China, extensive regional geological surveys, mineral prospecting activities, seismo-geological and geophysical explorations, and both specific and interdisciplinary research have provided a great abundance of basic geotectonic data. Moreover, the varied and complicated geological structures of the country have provided a favourable material base for the development of geotectonic theories, thus making geotectonics one of the most active research fields of geological sciences in the country. Geologists with different academic views have investigated the tectonic evolution of China; and their different methods of approach and varied geotectonic theories have gradually established different geotectonic schools of thought which in turn have given impetus to a general development of the study of geosciences.

From the 1950's there have appeared several relatively comprehensive synthesizing research works on the geotectonics of China. They are the *Geotectonic Map of China and Areas of Neighbouring Countries*, scale 1: 4,000,000, with an explanatory note entitled *An Outline of the Geotectonics of China* compiled by the Institute of Geology, Academia Sinica (Zhang Wenyou as Editor-in-Chief) (1959), the *Geotectonic Map of the People's Republic of China*, scale 1:3,000,000, with an explanatory note entitled *The Main Characteristics of the Geotectonics of China* compiled by Huang Jiqing as Editor-in-Chief (1962)[1], *The Tectonic System Map of the People's Republic of China*, scale 1:4,000,000, compiled by the Institute of Geomechanics (1976), and the *Tectonic Map of China*, scale 1:4,000,000, with an explanatory note entitled *An Outline of the Tectonic Characteristics of China* compiled by the Guangzhou Seismological Party of the State Seismological Bureau under the direction of Chen Guoda (1977). In addition, a series of scientific papers have been published by Yu Deyuan (1954), Zhang Bosheng (1962), Ma Xingyuan (1961), Wang Hongzhen (1955) and others, making valuable contributions to geotectonic studies in China. In recent years, Li Chunyu (C. Y. Lee) (1978), Guo Lingzhi and others (1980) have begun attacking the problem of geotectonics in China from the viewpoint of plate tectonics.

The newly compiled 1:4,000,000-scale *Tectonic Map of China* represents the results of a scientific research project accomplished collectively by the Section of the Structural Geology of the Institute of Geology, Chinese Aca-

---

1 Unpublished data.

demy of Geological Sciences, under the direction of Professor Huang Jiqing (1979) and with the vigorous support of geoscientists from various institutions and regions of the country. The map incorporates as many geological and geophysical data as possible (mainly unpublished data from regional geological and seismogeological investigations, petroleum prospectings, geophysical explorations, etc.) accumulated during the period from the publication of the 1:3,000,000-scale *Geotectonic Map of the People's Republic of China* in 1960 up to mid-1978. The map also gives a detailed classification of the first- and second-order tectonic units and main fracture systems in China, based on field researches in some regions together with analysis of Landsat photographs. Moreover, the map marks out plate tectonic indications such as ophiolite, mélange and high-pressure and low-temperature metamorphic belts (glaucophane-schist belts, etc.) so far discovered, outlines the isobaths of the Moho within the Chinese territory, and depicts with special attention the origin, development, intersection and compounding relationship of the three major tectonic domains of China, namely the Pal-Asian, the Marginal-Pacific and the Tethys-Himalayan.

The preparation for the compilation of the 1:4,000,000-scale *Tectonic Map of China* began in 1974 and was in full swing in 1975. With Ren Jishun as chief organizer, the compilers of the map divided the work among themselves, with Liu Xun being responsible for the section *the Sino-Korean Paraplatform;* Tang Yaoqing, for the section the Yangtze *Paraplatform;* Chen Bingwei and Ai Changxing, for the section *the South China Fold System,* including the Hainan Island; Jiang Chunfa, for the section *Western Segment of the Tianshan-Hinggan Geosynclinal Fold Region and the Tarim Platform;* Zhang Qinwen and Xu Zhiqin, for the section *Eastern Segment of the Tianshan-Hinggan Geosynclinal Fold Region,* that is, the Northeast China region; Xiao Xuchang, Qu Jingchuan and Zhu Zhizhi, for the section *Kunlun-Qinling Geosynclinal Fold Region;* Qin Deyu and Li Guangqin, for the section *Yunnan-Tibet and Himalayan Geosynclinal Fold Region;* Zhang Zhengkun, for the section *Taiwan Fold System;* Ren Jishun, together with Xu Zhiqin, Liu Xun and Zhang Qinwen, for the study of the material on the Meso-Cenozoic tectonics in eastern China and for the collection of data on deep-seated structures. By the end of 1975, in preparation for the 25th International Geological Congress, the compilation of a 1:10,000,000-scale *Geotectonic Map of China,* (published in 1976) was accomplished as the first step for the compilation of a 1:4,000,000-scale geotectonic map, with a brief explanation entitled *An Outline of the Geotectonic Characteristics of China* (Huang Jiqing et al. 1977). The compilation work of the 1:4,000,000-scale map suspended in 1976 when most of the compilers were undertaking geological prospecting for iron ore deposits. Then in 1977, Ren Jishun, Jiang Chunfa, Zhang Zhengkun, Qin Deyu and others resumed compilation with Ren Jishun as leader. The division of work was as follows: Jiang Chunfa was responsible for the section *Tianshan-Hinggan Geosynclinal Fold Region and the Tarim Platform;* Zhang Zhengkun, for the section *the Sino-Korean Paraplatform and the Taiwan Fold System;* Qin Deyu, for the section *the Kunlun-Qinling and Yunnan-Tibet Geosynclinal Fold Regions and the Hima-*

layan Fold System (with Cheng Guoming participating in work on the Tibet Region); Ren Jishun, for the section *the Yangtze Paraplatform and the South China Fold System;* and Xie Liangzhen, for cartographic designing. In the Autumn of 1978, the draft map was completed and the cartographic drawing was made by Xie Liangzhen and Shen Yonghui. At the end of the same year the finished drawing was delivered to the Cartographic Publishing House for publication.

The present monograph is both a brief explanatory note to the 1:4,000,000-scale *Geotectonic Map of China,* and a sequel or further elucidation of the work *An Outline of the Tectonic Characteristics of China* (Huang Jiqing et al. 1977). In this monograph the authors underline methodologically the need of an integrated study of all tectonic structures, continental and oceanic whether shallow-seated and deep-seated, regional and global, and microscopic and macroscopic, of combining quantitative and qualitative studies, and of optimal use of new techniques and methods. Theoretically, the monograph places emphasis on mobile belts and stable regions and their transformation, with special attention to deep fractures from the viewpoint of deep-seated structures, gives a new classification of essential structural types of crust and upper mantle, with due consideration of some important research results in plate tectonics; and finally puts forward a concept of continent-marginal activization belt. In the specific description of the geotectonics of China, based on the abundant practical data accumulated by broad ranks of Chinese geologists since the founding of new China and the results of field investigations and thematic research in South China, Sichuan, Yunnan, Qinling Mts., Qinghai, Tibet, Xinjiang and other regions, the authors have systematically classified the tectonic cycles and tectonic units of China, elucidated with emphasis the geosynclines and deep fractures, and established a model for the tectonic development of China.

This monograph is written by Ren Jishun, Jiang Chunfa, Zhang Zhengkun and Qin Deyu under the direction of Prof. Huang Jiqing. Chapters II, III are a joint work of the four authors, with the actual preparation of Chapter I by Ren Jishun, of Chapter IV by Jiang Chunfa and Ren Jishun, of Chapter V by Ren Jishun and Qin Deyu, and of Chapter VI by Ren Jishun and Zhang Zhengkun. Professor Li Chunyu read the completed manuscript.

In compiling the 1:4,000,000-scale *Geotectonic Map of China* and the present monograph, the authors have received vigorous support and assistance from various geological bureaus, research institutes and regional survey parties from all provinces, municipalities and autonomous regions; all regional research institutes under the Chinese Academy of Geological Sciences; the Institute of Geophysical and Geochemical Exploration, the Airborne Geophysical and Aerogeological Survey Division, the Integrated Research Party of Petroleum Geology and the Jiangsu Central Laboratory for Petroleum Geo- logy under the former State Bureau of Geology (now renamed the Ministry of Geology and Mineral Resources); the State Seismological Bureau and its Institute of Geophysics; the Institute of Geophysics of the Chinese Academy of Sciences (*Academia Sinica*); the Department of Geology and Geography,

Beijing University; and the Department of Geology, Nanjing University.

The authors take this opportunity to express their deep-felt gratitude to the above-mentioned institutions and organizations. The authors also acknowledge with thanks the energetic collaboration and enthusiastic support given by their colleagues from all sections of the Institute of Geology, Chinese Academy of Geological Sciences, and wish to extend their sincere appreciation to the Map Publishing House for final drawing and publication of the map, to the China Geological Map Printing House for the printing of the map, and to the Science Press for the printing and publication of this monograph.

Special thanks are extended to Zhou Yaoxiu and Cheng Zhengyan for the contribution of their latest, still unpublished research results, including the Bouguer Gràvity Anomaly Map of China, the Moho Depth Map and several gravity profiles; and to Guo Lingzhi, Shi Yangshen, Wang Henian and Wu Jiashan for providing the authors with several copies of very valuable photos.

The authors gratefully acknowledge the drawing of the textfigures by Weng Huihua and Dong Xiaojing.

The boundary of China in all the attached maps in this book is drawn after the *Map of the People's Republic of China* published by the Cartographic Publishing House in 1980.

. Under the supervision of Professor Huang Jiqing (T. K. Huang) the translation of the book from Chinese into English was done by:

Jiang Minxi (Preface and Chapter I);

Xu Niansheng (Chapters II and VI);

Fei Zhenbi (Chapter III);

Wang Ting (Chapter IV, also collated and verified the English translation of Chapter V); and

Liu Mingquan (Chapter V. also collated and verified the whole English version of the book except for that of Chapter V).

Section of Structural Geology,

Institute of Geology,

Chinese Academy of Geological

Sciences.

Contents

Chapter I

On the Problem of Methods and Theories

I.1 ON THE METHOD OF HISTORICAL ANALYSIS

All the matter forming the Earth's crust and mantle is in a process of constant motion, change and development. In simple words, the so-called method of historical analysis is a method which is used to probe, according to the order of historical evolution of the Earth and on the basis of all available geological, geochemical and geophysical data, into the salient features of geotectonic development in different stages of geological history; and on the basis of investigation and comparison of the origin, development and transformation of the various parts of the crust and mantle, to find out their general character as well as their differences and thus to elucidate the laws governing their movements. All motion is, without exception, a motion of matter. There is no motion that is divorced from matter, and there is no matter that is not in motion. Therefore, the study of geotectonics must be based on the study of the material composition of the crust and the mantle, i.e. the constituents, properties, types and distribution of rock formations, as well as all their changes in time and space during the evolution of the crust and the mantle, including the changes in the material composition and structural forms. In broad outline, such a study can be summarized as involving sedimentary formations, magmatism, tectonism, metamorphism, and metallogeny, as well as the integrated analysis of geophysical (gravity, magnetic, electric and seismic surveys, heat-flow measurements, etc.) and geochemical data.

All tectonic phenomena manifested at the Earth's surface were formed in different geological times and originated from different depths. They are interrelated with one another, constituting an organic whole. In the past, the study of geotectonics was mostly carried out on land areas with particular attention to the analysis of surface and regional geological structures. In recent years, however, along with the development of disciplines such as marine geology, geophysics, geochemistry, isotope geology and experimental geology, and the extensive application of new techniques and methods such as space navigation and remote sensing, the mutual penetration and integration of various disciplines and the emergence of new frontier sciences have steadily expanded the realm of geological sciences from continent to ocean, from surface to greater depth of the Earth, from selected regions to the entire globe, and from qualitative to quantitative analysis. Hence, it has become possible to integrate the investigations of continental, surface, regional, and microscopic structures respectively with the investigations of oceanic, deep-seated, global and macroscopic structures, and to link qualitative with quantitative investigation. Furthermore, the development of cosmonautic techniques has enabled a comparative research of the Earth with that of other planets and a close link between the study of the Earth's structure and that of the cosmic ones, thus allowing geoscientists to observe and investigate the Earth's structure as a whole from outer space. Consequently, a fundamental change has taken place in the process of cognition of the Earth's structure: the cognition process in the past was a point→line→plane process, i.e., a process from the part to the whole, whereas the cognit on at present proceeds from the whole to the part, with a close integration of the two.

I.2   ON SOME THEORETICAL PROBLEMS OF GEOTECTONICS

In the past 10 to 20 years, the theory of geotectonics has developed by leaps and bounds owing to the advance and great diversity of research methods, and the expansion and deepening of realms of research. Most important in this respect is the emergence and development of the concepts of sea-floor spreading and plate tectonics. Moreover, following the steady progress in the study of global tectonics, geologists have been paying growing attention to the integrated research of geotectonics with such geological processes as sedimentation, paleogeography, magmatism, metamorphism and tectonic disturbance. This has not only vigorously stimulated the development of geotectonics, but also led to a more deep-going study in all branches of geology.

On the basis of the long-period studies of geotectonics in China, with a due consideration of the latest research results available both from abroad and at home, the authors will give a brief discussion on certain theories and concepts, some of which have already been proved by facts, while others are still to be verified by practice in future.

I.2.1   Continents, Oceans and Their Contact Relationship

Recent geological and geophysical data have fully demonstrated that two different types of crust-upper mantle structures exist: the continental type and the oceanic type. In the case of the continental type of structure the Earth's crust is roughly split into two layers, i.e. the upper sial and the lower sima, with a general thickness of 35—40 km and over 70 km maximum, and with the low-velocity layer of the upper mantle being at a lower position, commonly at 120 km or even deeper below the surface (Zeng Rongsheng, 1973). For the oceanic type structure, however, sedimentary formations are underlain immediately by the simatic layer, with complete absence of the sialic layer. Therefore, the Earth's crust is comparatively thin (usually 5—8 km thick) with the low-velocity layer of the upper mantle being at a higher position, generally at a depth of 60 km and more (Zeng Rongsheng, 1973). This, of course, is only a rough and generalized scheme of classification. Practical observations have indicated that either the Earth's crust or upper mantle is multi-layered, and both are composed of masses of alternated layers with different seismic-wave velocities. Figure 1 clearly shows the model for the multilayered crust under the North China Plain.

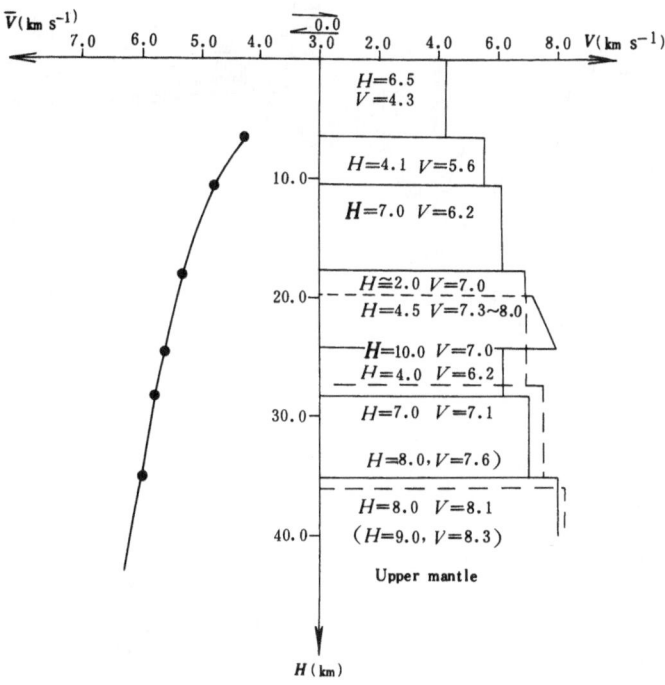

Fig.1. The velocity distribution in the media of the Earth's crust and the top part of the upper mantle and the model for the Earth's crust under the North China Plain.

(After Teng Jiwen et al. 1974)

Two basic types of contact relationship between continents and oceans have been distinguished: the Atlantic type and the Pacific type. The Atlantic type of contact relationship is characterized by its tensional nature, smaller activity, relative stability of the continent contiguous to the ocean, and the occurrence of tensional and fault-block structures on the continental margin. The contact relationship between the Atlantic Ocean and the continents of North and South America, of Africa, and of Europe is of this type, and thus it is so named. Meanwhile, the Pacific type, being compressional and/or shear-compressional, is characterized by stronger activity of its adjacent continents, the presence of gigantic deep fracture zone (the Benioff Zone) incising mantle on continental margins, and the occurrence of deep trenches, of Andes-type coastal ranges or island arcs, of marginal seas, as well as of zones

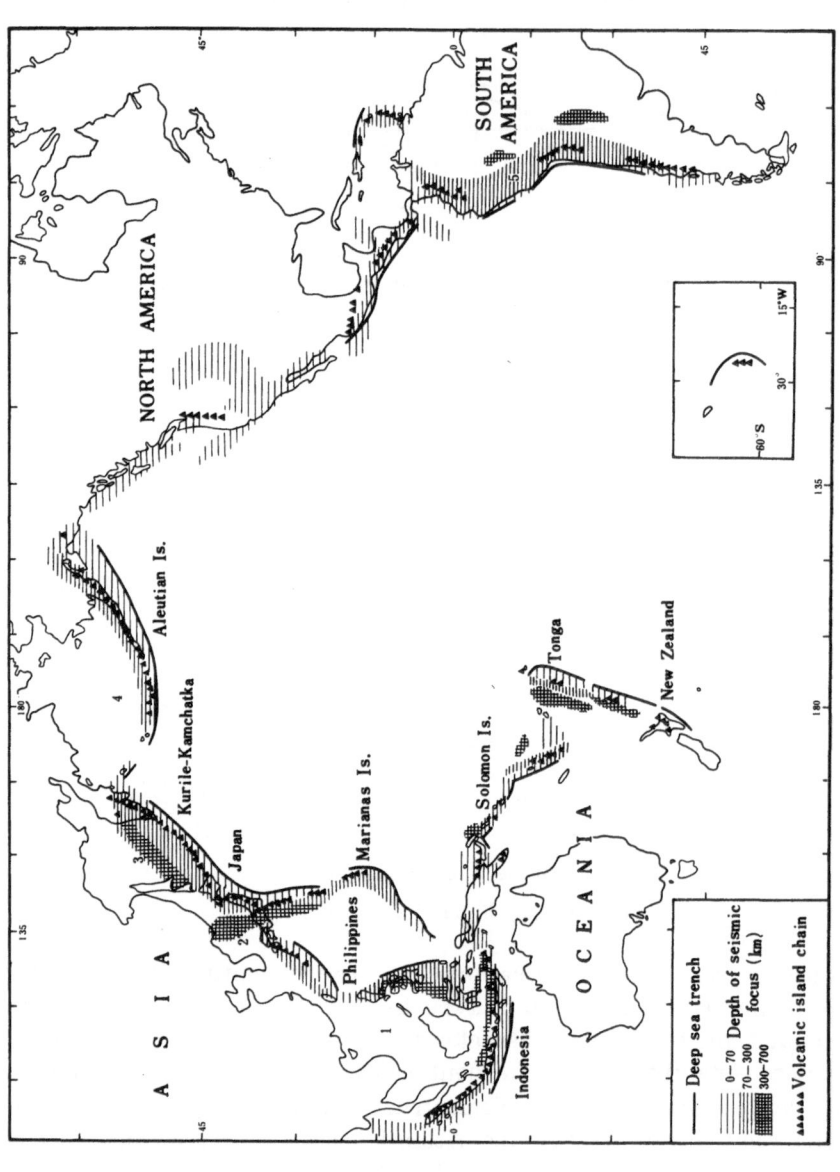

Fig.2. Active continental margins around the Pacific Ocean. (After Hatherto 1974)

1 South China Sea; *2* the Sea of Japan; *3* the Sea of Okhotsk;
4 the Bering Sea; *5* the Andes

Deep sea trench

Depth of seismic focus (km)
0—70
70—300
300—700

Volcanic island chain

NORTH AMERICA

SOUTH AMERICA

ASIA

OCEANIA

Aleutian Is.

Kurile-Kamchatka

Japan

Philippines

Marianas Is.

Solomon Is.

Tonga

New Zealand

Indonesia

of intense earthquake and volcanic activity (Fig. 2). Such a contact relationship is called the Pacific type because the Pacific Ocean is found to be in the described pattern of relations with the continents of Asia, South and North America, and Oceania. As was pointed out by Mitchell and Reading (1969), the above-mentioned two types of contact relationships, though different from each other, are mutually transformable.

I.2.2 Mobile Belts, Stable Regions and Their Transformation

Mobile belts and stable regions are both objective entities that exist in nature. The salient features of geosynclines as mobile belts, and of platforms as stable regions have already been described in sufficient detail in the geological literature. Specifically speaking, island areas are furnished with intense volcanic activity, folding and faulting, and represent mobile belts, while shields, e.g., the Canadian shield, are stable regions, where pre-Paleozoic crystalline basement crops out on a peneplained surface with only a thin (if any) sedimentary cover.

From the viewpoint of deep-seated structures, we hold that the essence of mobile belts lies in the heterogeneity of the constituents both of the Earth's crust and upper mantle, the great intensity of the motion of matter, the general presence of low-velocity layers in the Earth's crust, a transitional relationship between crust and mantle, obscure Moho discontinuity, existence

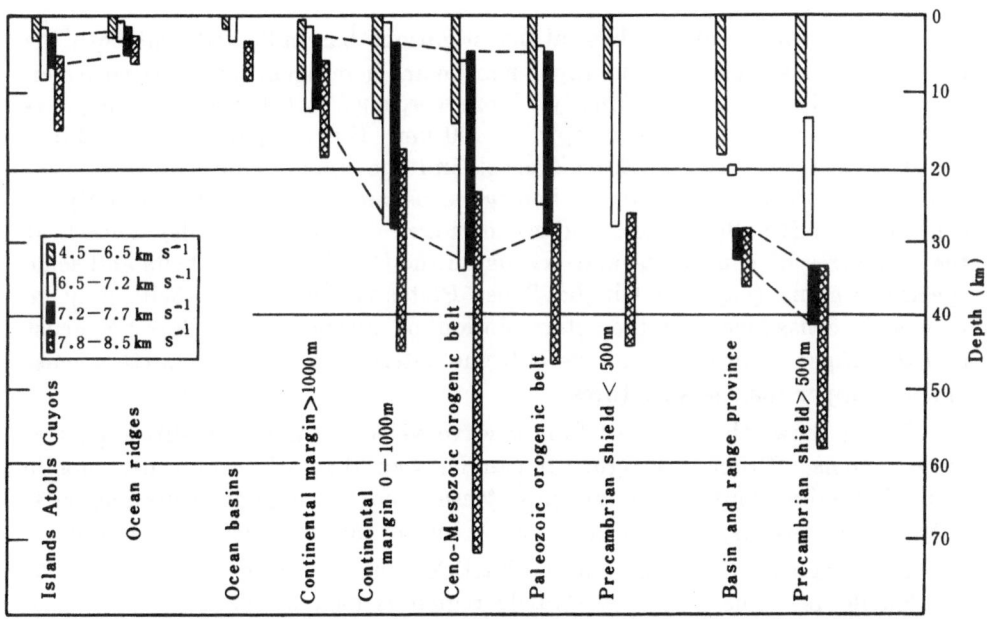

Fig.3. Depths of four ranges of P-wave velocity in different tectonic regions

(a synthesis of world data). (After Drake et al. 1968)

The four velocity ranges are corresponding to: 4.5—6.5 km s^{-1}, upper crust; 6.5—7.2 km s^{-1}, stable lower crust; 7.2—7.7 km s^{-1}, abnormal crust or mantle in tectonically active regions; 7.8—8.5 km s^{-1}, upper mantle

of an anomalous interval with a P-wave velocity of 7.2—7.7 km s^{-1} between crust and mantle, and a small depth and great thickness of the low-velocity layer in the upper mantle. The stable regions, however, are characterized by a relative homogeneity of constituents in crust and upper mantle, slow motion of matter, absence of anomalous interval on the top of upper mantle, sharp contact between crust and upper mantle, and a great depth and small (or even unnoticeable) thickness of the upper mantle's low-velocity layer.

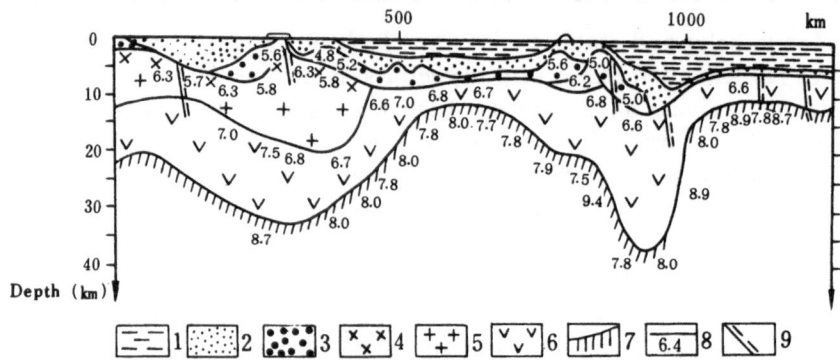

Fig.4. A section of the crust from Sakhalin to Kurile Island. (After Tujezov 1975)

1 water; *2* inferred Cenozoic sediments; *3* inferred Mesozoic sediments; *4,5* granites (*4* assumed to be of Paleozoic and Precambrian age; *5* of unknown age); *6* basalts; *7* the Moho; *8* P-wave velocity, km s^{-1}; *9* deep fracture

A large volume of geophysical investigations has indicated that both the modern Earth's crust and the upper mantle are heterogeneous in structure either longitudinally or latitudinally. Moreover, such a heterogeneity may extend downwards at least to a depth of 400 km. But comparatively speaking, the structures of both the Earth's crust and the upper mantle in most continents and oceans are relatively homogeneous and stable, and it is only in continental rift-valley zones, mid-ocean ridges (oceanic rift-valley zones), at the junctures of continents and oceans, in the Cenozoic folded belts and some specific regions (e.g., the Qinghai-Tibet Plateau) that mobile belts may be formed in consequence of the active motion of matter in the Earth's crust and the upper mantle, and of the extreme heterogeneity of the Earth's crust and the upper mantle structures.

Fig. 3 shows the depth of four P-wave velocity ranges in different tectonic regions. From this figure it can be seen that all tectonically mobile belts, including continental margins, Meso-Cenozoic orogenic belts and rift-valley zones, are commonly characterized by an abnormal crust or mantle.

Fig. 4 is a profile section of the Earth's crust from the Sakhalin Is. to the Kurile Is. The Sakhalin—Kurile region is part of the developing geosynclines, that involve eastern Asia and the western margin of the Pacific Ocean. This figure shows that both the Earth's crust and upper mantle are extremely inhomogeneous in structure.

It is evident from Fig. 5 that the Earth's crust shows a variation in structure from extreme heterogeneity to relative homogeneity along a direc-

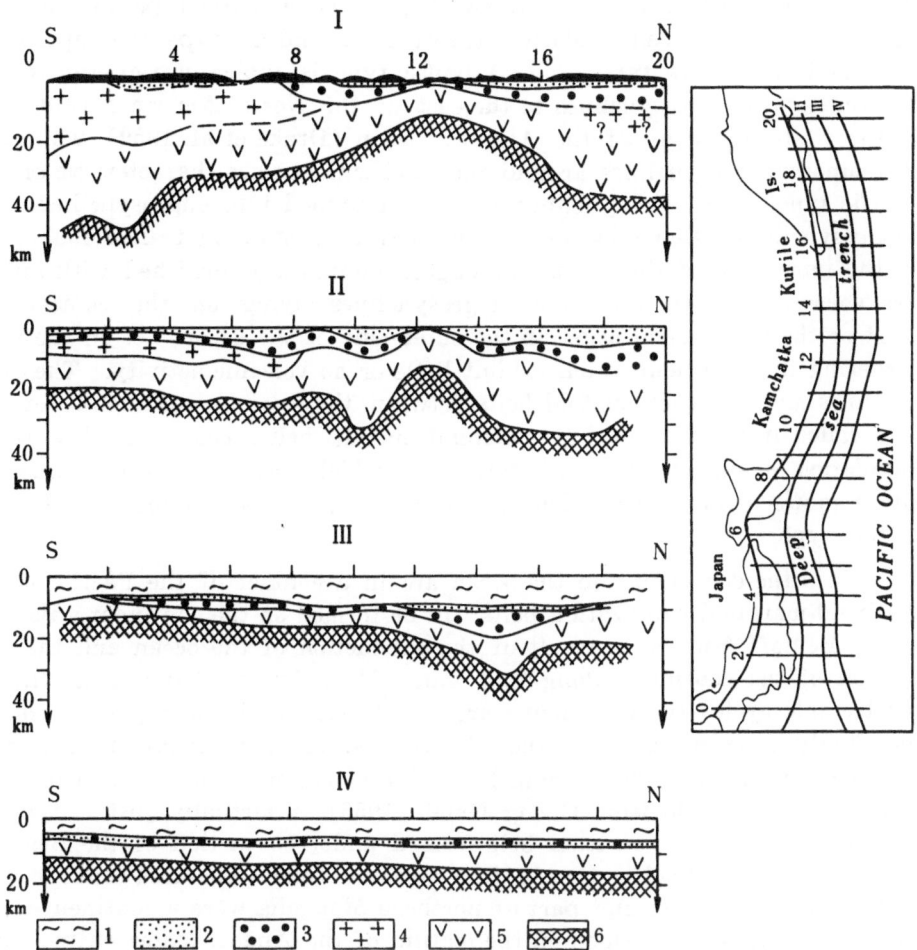

Fig.5. Diagrams showing variations in the structure of the Earth's crust from Japan–Kurile–Kamchatka Island Arc to the Pacific Ocean. (After Tujezov 1975)

1 water;   *2,3* sedimentary rocks (*2* upper layer;   *3* lower layer);   *4* granites;
5 basalts;   *6* top part of upper mantle. *I* Japan-Kurile-Kamchatka Island Arc; *II* slope of the Japan-Kurile-Kamchatka Island Arc adjacent to the Pacific Ocean; *III* Japan-Kurile-Kamchatka deep trench; *IV* floor of the Pacific Ocean close to Japan and Kurile

tion from the island arc (contemporaneous geosynclinal zone) towards the Pacific Ocean (oceanic platform).

From the viewpoint of modern geology, the oceanic rift zone is a region where eugeosyncline originates. It is in such a region that ophiolite suites occur. When an oceanic rift zone is developed into an ocean basin, oceanic marginal mobile zones will be formed in areas adjacent to continent. They are the geosynclinal belts and geosynclinal fold belts referred to by many geologists. Corresponding to the two different types of the contact relationship between the continent and the ocean, geosynclines located on the oceanic margins can also be classified into two types: the Atlantic type and the Pacific type, the latter being subdivided into the island arc type, the Japan-Sea type, and the Andes type. The Atlantic type is typically represented by the contemporary geosynclines to the east of the North American continent on the western margin of the Atlantic Ocean (Drake et al. 1959), while the contemporary geosynclines around the Pacific Ocean are characteristic of the Pacific type. These geosynclines can be subdivided into eugeosynclines and miogeosynclines. Being located on the oceanic crust or the transitional crust on the inner side of the ocean, the eugeosynclines are furnished with intensive volcanic activity, while the miogeosynclines, lying on the continental crust or the transitional crust on the outer side of the ocean, i.e. on the side bordering the continent, often exhibit little or no volcanic activity. The geosynclines and geosynclinal fold belts between the various continents, otherwise called intercontinental geosynclinal mobile belts, commonly lay their foundation on immature oceanic rift zones which have not been developed into gigantic ocean basins. The northern Qilian Eugeosyncline may belong to this type.

When the continent and the ocean are in a contact of the Pacific type, continent-marginal activization belts will be formed on the continent adjacent to the Benioff Zone as a result of the subduction of the ocean and the obduction of the continent along that zone. It is obvious that during different tectonic cycles the continent-marginal activization belts may form correspondingly in accordance with the relative position and interrelation of the continent and ocean. The so-called reactivized eastern China platform suggested by some geologists (Cheng Guoda 1956) is virtually just a typical continent-marginal activization belt formed under the action of the Benioff Zone in the West Pacific Ocean since Mesozoic time. The area south of the Baikal Lake in Siberia and part of northern Mongolia were a continent-marginal activization belt formed on the margin of the Angara paleocontinent in the Caledonian and Variscan stages. The Qinghai-Tibet Plateau-type uplifts, the Tianshan-type rejuvenated mountain systems, and the related piedmont and intermontane depressions were formed when the subduction along the Benioff Zone culminated with the disappearance of the oceanic crust, the closing of the ocean and the collision between two large pieces of adjoining land masses. These should belong to another type of continent-marginal activization belts which appear to be within the continent and differ in characteristics from the eastern China continent-marginal activization belt.

To sum up, the crust-upper mantle structure may be classified as follows:

(1) *Stable regions,* which are subdivided into:

a) *Continental stable regions*: These are platforms characterized by the presence of a continental crust, as represented by the Russian Platform, the North American Platform and the Sino-Korean Paraplatform; and

b) *Oceanic stable regions* (or oceanic basins): These are platforms with an oceanic crust. Generally, this is only a tectonic unit in the Earth's history. For instance, the West Pacific Ocean is simply an oceanic platform developed since Mesozoic. The earlier oceanic platforms have vanished due to the closing of the ocean.

(2) *Mobile belts,* which are subdivided into:

a) *Oceanic mobile belts*: These are geosynclinal mobile belts which embrace oceanic rift mobile belts and oceanic marginal mobile belts, the latter being divisible into the Atlantic and the Pacific types. Typical eugeosynclinal belts (with ophiolites) generally originated from the oceanic rift belts and have undergone a process of development from an oceanic rift mobile belt to an oceanic marginal mobile belt. Intercontinental geosynclinal mobile belts might form when the oceanic rift is not well developed.

b) *Continental mobile belts*: These comprise both continental rift belts and continent-marginal activization belts. For example, eastern China is a huge Meso-Cenozoic continent-marginal belt.

Here, the problem of tectonic activization needs to be dealt with in more detail.

Tectonic activization phenomenon was first noted by Wong (1926). In 1940, Mirchink suggested that this was a tectonic unit of the third type, differing from both the geosyncline and the platform. Since then, such tectonic phenomena have been studied by more geologists and referred to as "activized region" or "post-platform orogeny" (Chen Guoda, 1959; Beloussov, 1964). Based on the theory of platform activization, Chen Guoda conceived the theory of Diwa (geodepression) at the end of the 1950's (Chen Guoda 1959).

Judging from the concept of modern geology, in accordance with the structural classification of the crust-upper mantle, we are of the opinion that the so-called activized region or Diwa is not a tectonic unit characteristic of a specific stage in the progressive process of mobilization-fixation-transformation, as expressed by the formula: $X \rightarrow$ geosyncline $\rightarrow$ platform $\rightarrow$ Diwa $\rightarrow Y$ of the crustal structure, but a tectonic unit that can be formed at any tectonic cycle alongside the geosynclinal mobile belt. The so-called activized region or Diwa should be largely assigned to the continent-marginal activization belt (e.g. in eastern China) and in some cases it should be regarded as a continental rift belt (e.g. East-African rift valleys which are referred to by Cheng Guoda as East-African Diwa), while others ought to be considered as rejuvenated geosynclines (e.g. the Youjiang River area).

Continental rift valleys may develop further into oceanic rift valleys. Oceanic marginal mobile belts, geosynclinal belts and geosynclinal fold belts are found in those parts of the ocean that adjoin the continent after the oceanic rift valleys have developed into oceanic basins (or oceanic platforms). When the continent and the ocean are in a Pacific-type contact, continent-marginal activization belts may be formed on the margin of the continent. When the continent and the ocean are in an Atlantic-type contact, the continent appears tectonically to be in the state of a more or less stable platform. Such is the generalized interrelation between the above-mentioned different tectonic types developed since the Paleozoic.

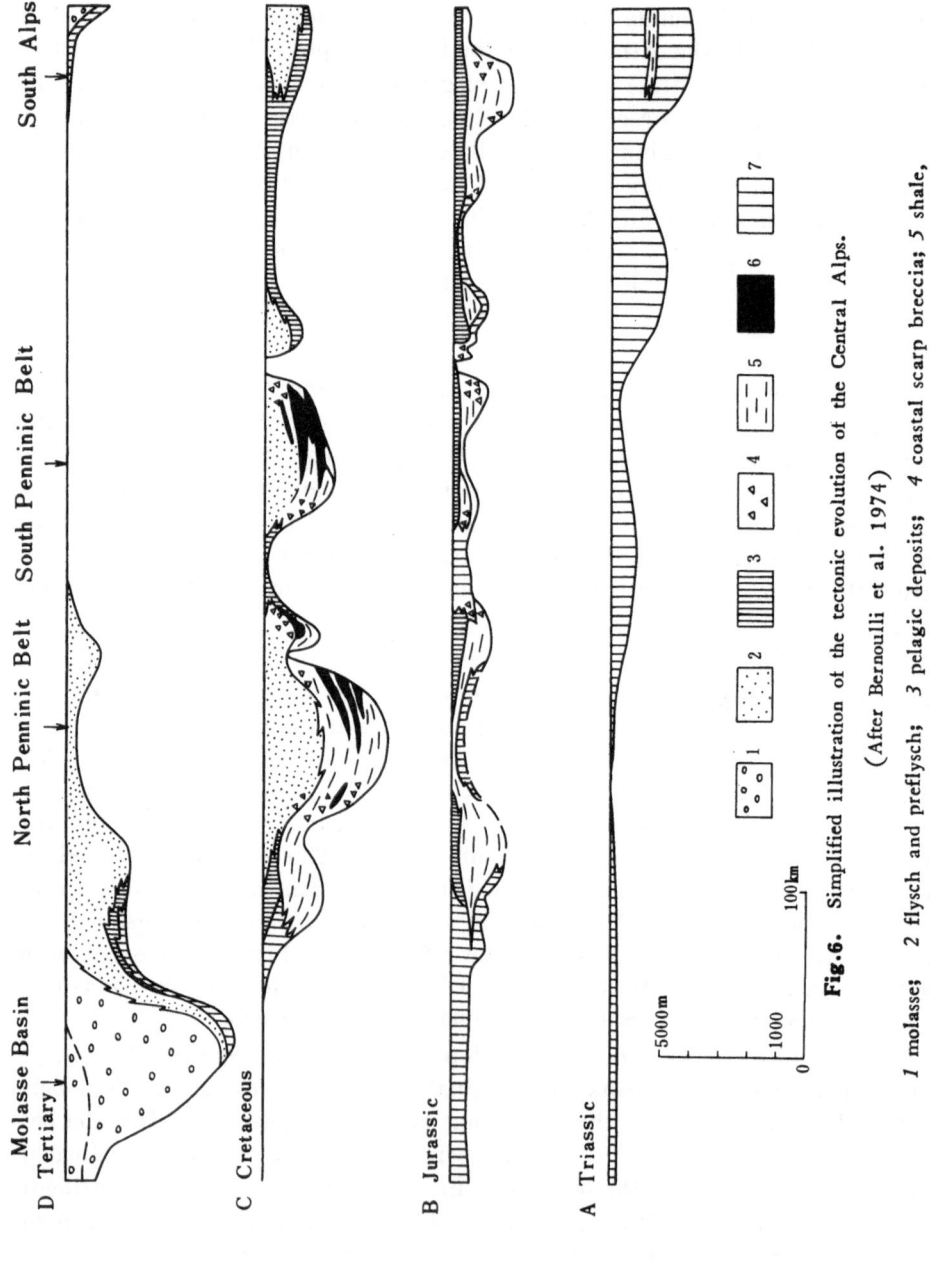

Fig. 6. Simplified illustration of the tectonic evolution of the Central Alps.

(After Bernoulli et al. 1974)

1 molasse; *2* flysch and preflysch; *3* pelagic deposits; *4* coastal scarp breccia; *5* shale, including schistes lustrés; *6* ophiolite; *7* shallow-water deposits, mainly carbonate rocks

Mobile belts and stable regions may be mutually transformable. The process of transformation from a mobile belt into a stable region is essentially a process of increasing homogeneity of the structure of the Earth's crust and mantle——a process of gradual change of the active-to-slow motion of matter. Conversely, the process of transformation of a stable region into a mobile belt is that of a change of the structure of the Earth's crust and mantle from homogeneity to heterogeneity——a process of gradual change of the slow-to-active motion of matter.

The Atlantic-type geosynclines and the Pacific-type geosynclines might be mutually transformable, and so are the geosynclines and the platforms, as well as the continents and the oceans. Examples for all these situations are

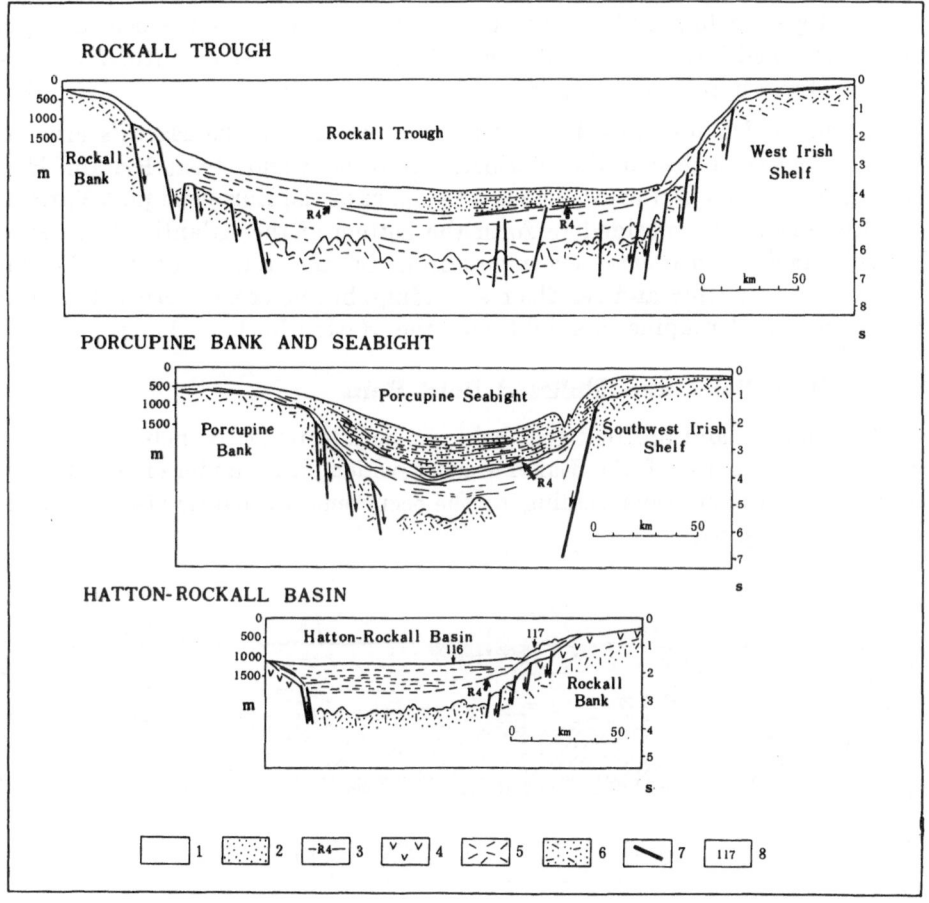

Fig.7. Correlation of the crustal structure in Rockall trough, Porcupine Seabight and Hatton-Rockall Basin. (After Roberts 1974)

1 floe-ice deposits; *2* terrigenous deposits; *3* Eocene; *4* Paleocene volcanics; *5* oceanic crust; *6* Precambrian-Carboniferous; *7* fault; *8* DSDP drillhole

quite common. The process of transformation of a platform into a geosyncline is virtually a process of transformation of a continental crust into an oceanic crust. This may be clearly explained by the formation and development of the Qilian Geosyncline in China and the Alpine Geosyncline in Europe.

Figure 6 is a simplified illustration of the tectonic evolution of the Central Alps, a well-known typical geosynclinal fold belt. During the Triassic it was still in a platform state, and a platform cover mainly composed of shallow-marine carbonate rocks was deposited on the basement of the post-Variscan platform. In Jurassic-Cretaceous times, owing to taphrogenesis, the continental crust transformed into an oceanic crust and thus the platform changed into a geosyncline, giving rise to abyssal geosynclinal deposits (schistes lustrés) and ophiolites. In the Tertiary, flysch and later molasse were deposited, and geosynclinal sedimentation came to an end, thus the oceanic crust was transformed into continental crust. Here we see that a continental crust was transformed into an oceanic crust and vice versa.

Figure 7 demonstrates three structural profiles of the Earth's crust of the Atlantic Ocean floor in the area between Iceland and Great Britain. Here the original continental crust, which was a part of the European post-Variscan platform, gradually subsided to form the bottom of the Atlantic Ocean since Cenozoic times. An oceanic crust appeared in the centre of the Rockall Trough due to rifting and sea-floor spreading, but no oceanic crust was formed yet in the Porcupine Seabight and the Hatton-Rockall Basin.

I.2.3 Rift Valleys and Uplifted Fold Belts

Rift valleys and uplifled fold belts are both tectonically mobile belts. Rift belts are those parts of the Earth's crust which have undergone extension and become thinner, thus leading to the occurrence of mantle rises and mantle pillows (Fig. 8).

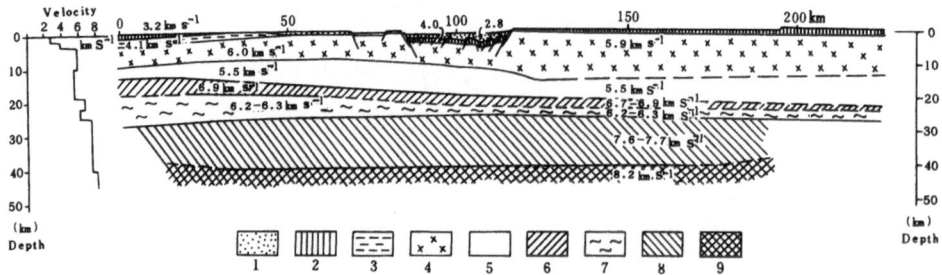

Fig.8. A profile across the Rhine Graben. (After Prodehl et al. 1976)

1 Quaternary and Tertiary; *2* Jurassic, Triassic and Permian; *3* Carboniferous and Devonian; *4* crystalline basement; *5* sialic low-velocity layer; *6* intermediate crustal layer; *7* intermediate low-velocity zone; *8* rift-valley soft "mattress" (mantle pillow); *9* upper mantle

Uplifted fold belts are parts of the Earth's crust where it has been subjected to compression and thus become thickened, resulting in the formation of mantle depressions and mountain roots. The formation of the mantle rises and depressions and the existence of transition between the Earth's crust and mantle demonstrate that the Moho discontinuity is by no means an unchangeable boundary, and that mutual transformation of matter has taken place time and again between the Earth's crust and the mantle[2]. In other words, the Earth's crust may transform into the mantle and vice versa. So far, the physicochemical mechanism for such a transformation, however, has not been explained.

Rift valley belts comprise continental and oceanic rift valleys, both being basically identical in the tectonic nature of the Earth's crust and the mantle, that is, there exists at depth a process of invasion of the Earth's crust by the mantle (the mantle's upward rising and invading to "devour" the Earth's crust). For this, the Red Sea may serve as an excellent example (Fig. 9).

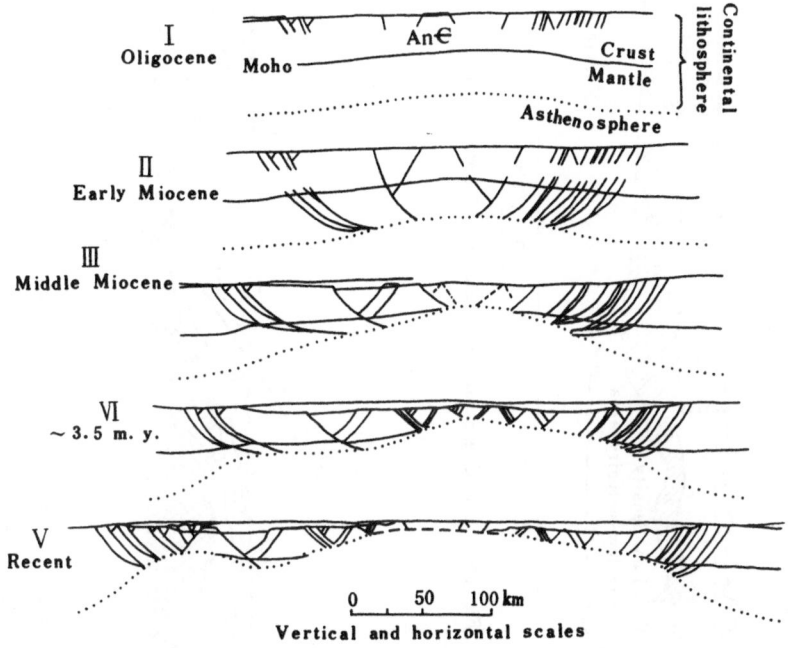

Fig. 9. Tectonic evolution of the Southern Red Sea.
(After Lowell et al. 1975)

The uplifted fold belts are divisable into geosynclinal ones and platformal ones (platformal folds) or block-uplift zones (regions). Geosynclinal up-

2 The authors acknowledge their appreciation to Xie Jize for the manuscript of his work *On the hypothesis of plate tectonics and other related problems*, in which one chapter deals with the transformation between the crust and the mantle.

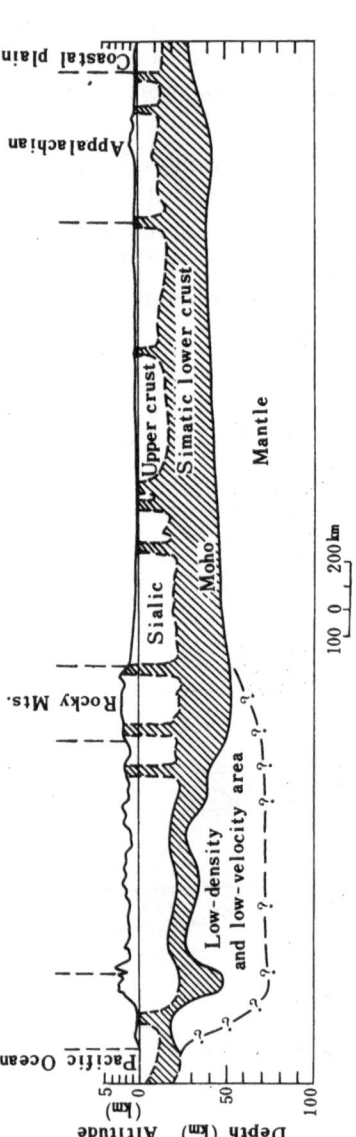

Fig.10. A profile across the continent of North America. (After Spencer 1977)

Fig.11. A simplified profile of structures of the Himalayas. (After Gansser 1966)

1 Precambrian shield (basement of Indian platform); *2* Gangdise granites; *3* ophiolite suites; *4* granitized rocks; *5* gneisses and migmatites; *6* Early Palaeozoic and older rocks; *7* Lower Palaeozoic and Upper Precambrian (Tethyan belt); *8* Mesozoic and Upper Paleozoic; *9* Tertiary.

MBF Main boundary fault; *CT* Central thrust

lifted foldbelts are products of the late stage of development of geosynclines, while the platformal folds and block-uplifts (such as the Mesozoic structures on the platforms in the eastern part of China) are products of the continental (marginal) mobile belts. The structures of crust and mantle of these two kinds of fold belts are basically the same in essence, that is to say, there may exist a process of transformation of mantle into crust, as in the western part of the U.S.A. (Fig. 10). In the area of the Qinghai-Tibet Plateau the same situations are found, as indicated by the clues available so far, where the commonly lower velocity of seismic waves for both the crust and mantle is likely to prove this theory. Unfortunately, there are few reports concerning the structures of the crust and mantle of the Tibet Plateau, except for some scattered relevant data published outside of the country.

When strong activities take place between gigantic tectonic units along deep fracture zones, the crustal structures on the margins of the originally more stable tectonic units will be intensively involved to form complicated fold-fracture belts, as exemplified by the overthrusts in the Himalayas and Longmen Mts. area, the former being the northern margin of the Indian Platform intensively involved in the Himalayan Fold, and the latter the northwestern margin of the Yangtze Platform. Previously, these were mistaken for geosynclinal fold belts (Fig. 11).

I.2.4 Tectonic Movements

From the point of view of geodynamics, tectonic movements are considered to include the following main types: compressional type, tensile type and shear type[3]. Folds, uplifts and subsequent angular unconformities represent one of the forms of expression of tectonic movements and are a reflection of compression (customarily referred to as folding or orogeny). Similarly, tensile fractures, depressions and intensified sedimentation are another form of expression of tectonic activities and a reflection of tension or taphrogeny. Both the former and the latter represent two different aspects of a unified process of tectonic movements. Comparatively speaking, tension (or taphrogeny) is mostly manifested by a prolonged progressive process of development, whereas compression (folding or orogeny) is commonly demonstrated by a short-term and sudden (or explosive) process of development.

Tectonic movements of the shear type are a form of tectonic movements often associated with compression and tension. The most important examples of the horizontal shear movements are the transform faults near the oceanic mid-ridges and the major strike-slip faults on land. Shearings associated with the Benioff Zone and major overthrusts are represented mainly by a vertical shear or an oblique shear.

In the course of the historical development of the Earth, compression, folding and uplifting in some regions have been inevitably accompanied in

---

3 Jiang Chunfa holds that there still exists a tortional type which is represented by folds and faults produced by a rotation tortional force caused by upward migration of the magma. Some ring-like structures are virtually related to such a tortional force.

space by tension and taphrogeny elsewhere, and the reverse is also true. Important examples of this are, for instance: after the Variscan cycle, the formation of Eurasia was accompanied by the gradual split of the Gondwana land; the strong folding and uplifting of western China and the formation of the famous mountain systems and the Qinghai-Tibet Plateau have coincided with the continuous extension and subsidence of eastern China and the adjoining marginal sea areas since the Himalayan cycle; and the formation of the Xingkai foldbelt and the disintegration of the Chinese Protoplatform both took place in the Early Paleozoic. Speaking of time, relatively slow and progressive tension or taphrogenesis often occurs in alternation with more or less abrupt or sudden compression (folding or orogeny), thus bringing

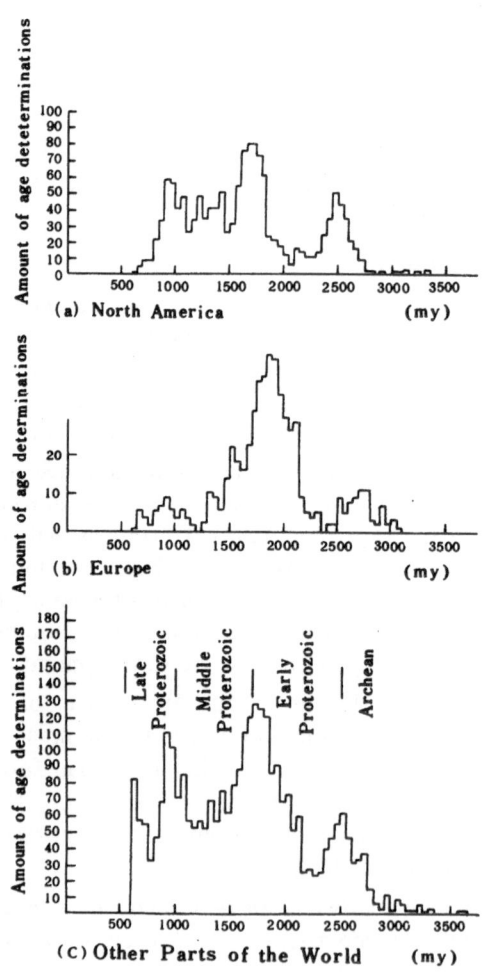

Fig. 12. A diagram showing frequency of distribution of isotopic ages of the Precambrian metamorphic rocks the world over.

(After Seyfert et al. 1973)

about successively developing tectonic cycles. Regional taphrogenesis conforms to the split of continents and the opening of oceans, but regional compression (folding or orogeny), as a rule, is intimately related to compression and collision between different crustal blocks (plates) of the Earth along the Benioff Zones (translithospheric fractures).

I.2.5 Monocyclicity and Polycyclicity

Although there is no absolute synchronism of global orogenic phases as suggested by Stille (1940), yet a relative coincidence in time of global tectono-magmatic cycles does exist in general terms. This is particularly true in one and the same tectonic belt. For instance, the Precambrian tectono-magmatic process obviously had three climaxes at about 2500 my, 1700 my, and 800 my respectively (Fig. 12). Other good examples are the Caledonian, Variscan and Alpine cycles, the Circum-Pacific Yanshanian Subcycle and the Cenozoic (or Himalayan) Subcycle in the Mediterranean, etc.

Stille established a monocyclic model for the tectono-magmatic development of geosynclines. Huang Jiqing proposed a polycyclic model for it instead (Fig. 13).

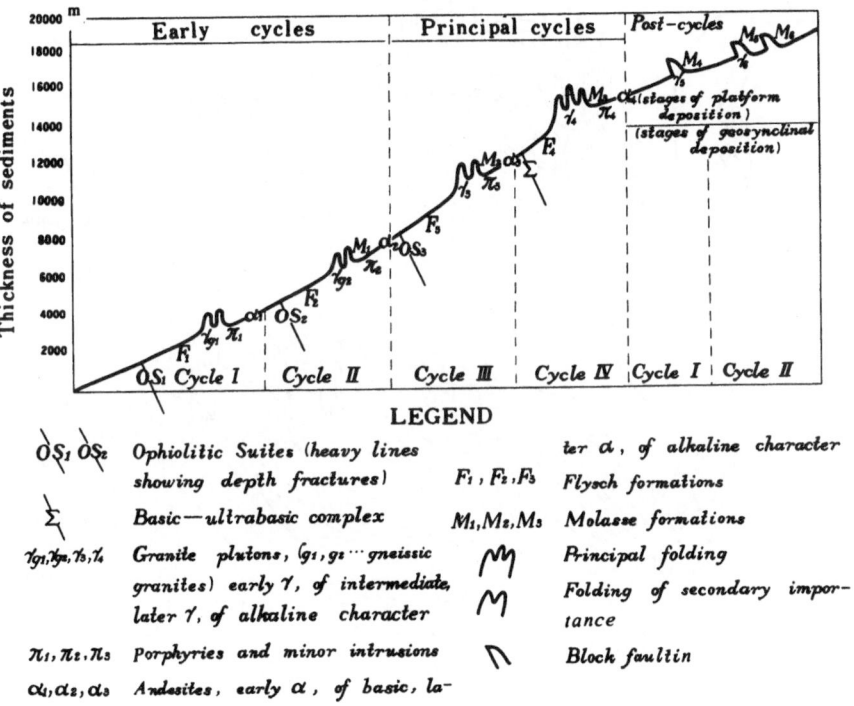

Fig.13. A sketch map showing polycyclic development of geosynclinal foldbelts.

(After Huang Jiqing 1979)

The present authors are convinced that the structure of the Earth's crust is developing polycyclically and in a spiral-like manner. The development of geosynclinal mobile belts is polycyclic, the development of the continent-marginal activization belts in eastern China is polycyclic, and all Pre-cambrian (or Pre-Sinian, to be more exact) platform basements the world over are, without exception, products of the development of polycyclic tectono-magmatism.

It can be seen from a tectonic sketch (Fig. 14) of the basement of the North American platform that the fold belts of different ages within the platform basement are not concentrically developed, but the older fold belts of different ages are very often cut across by the younger ones. The same is generally true with the development of the geosynclinal fold belts since the Paleozoic. Hence, the conception of simple continental accretion should be regarded as invalid. What is practically true is that not only growth oc-curred but also destruction of the continent; that is, there was destruction of the continent in the process of its growth, as well as growth of continent during its destruction; continental growth took place in some places, but continental destruction in others; in some regions continental growth was the main trend in one period, while continental destruction predominated in another period.

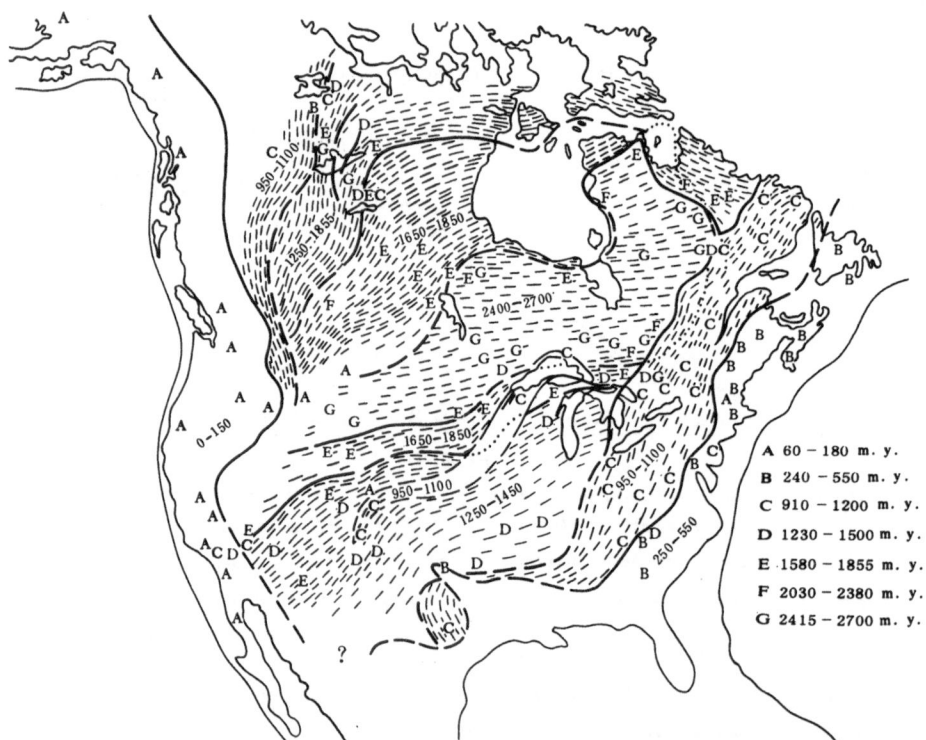

Fig.14. A sketch map showing the basement structure of the North American platform. (After Spencer 1977)

Hence, the so-called polycyclic development of the crustal structure im-plies not only the polycyclicity of tectonic movements but also the polycycli-city of sedimentation, magmatism, metamorphism and metallogeny that are closely related to tectonic movements. That is to say, the spiral-like polycy-clic development of the crustal structure denotes the summation of tectonic movements and various related geological processes, and it is essentially the inevitable result of repeated struggles and mutual transformation between the two basic tectonic units, the continent and the ocean.

It is thus obvious that the tectonic patterns and characteristics of different cycles must differ from one another to reflect various stages of tectonic evolution of the Earth's crust.

Based on a preliminary analysis of the data available both from abroad and at home, the authors tend to classify the tectonic cycles since Archaean as shown in the following table

Cycle	Duration (my)	Main tectonic events
Early Archean megacycle	>3000	Initial formation of the continental crust
Late Archean megacycle	3000—2500(±100)	Formation of the nuclei of the major platforms all over the world
Early Proterozoic megacycle	2500—1700(±100)	Formation of the major platforms
Late Proterozoic megacycle	1700—800(±100)	Final formation of the major platforms
Early Neogaic megacycle	800(±100)—230	
Xingkai cycle	800(±100)—540	
Caledonian cycle	540—400	
Variscan cycle	400—230	Formation of Laurasia at the end of the Variscan cycle
Late Neogaic megacycle	Since 230	Disintegration of Gondwana land since late Triassic; intensive activity of the Mediterranean tectonic belt and the Circum-Pacific tectonic belt; gradual formation of the modern continent-ocean pattern.
Apline cycle { Indosinian subcycle Yanshanian subcycle Himalayan subcycle		

I.2.6 Tectonic Transition and Tectonic Migration

Tectonic transition refers to the gradual transition between two different types of tectonic status, which is a common occurrence in nature.

Spatially, there is a transition between the Earth's crust and the mantle. This is shown in rift valleys and other structures (Fig. 15). There are also transitions between continents and oceans, e. g. the transition between the Asian continent and the Pacific Ocean, and between the American continent and the Atlantic Ocean. Transition also exists between mobile belts and stable regions as illustrated by that between the South China Caledonian Geosyncline and the Yangtze Paraplatform. Figure 15 shows that there exists, as a rule, a transitional layer between the Earth's crust and the mantle in every rift valley belt.

Figure 16 shows the tectonic transition between the Yangtze Paraplatform and the South China Caledonian Geosyncline.

Temporally, there is a transition in the transformation from geosynclines into platforms, but the transition in the transformation from platforms into geosynclines is more common. The former is represented by the tran-

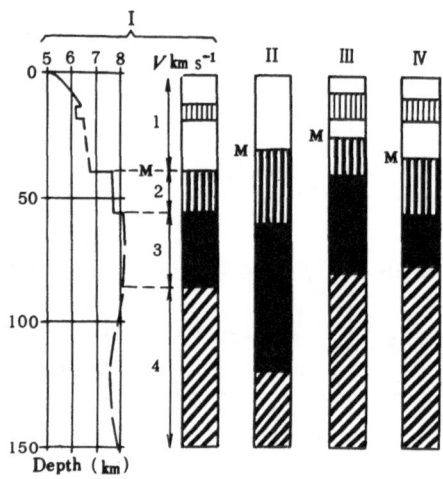

Fig.15. Columnar diagram showing the P-wave velocity in continental rift valley belts of the world. (After Puzyrev et al. 1978)

1 Earth's crust (with a low-velocity layer in it);   *2* mantle pillow (7.7km s^{-1}), crust-mantle transitional layer; *3* upper mantle (8.0—8.2 km s^{-1}); *4* low-velocity layer; *M* the Moho

I the Baikal Rift Valley;   *II* the northern part of the East-African Rift Valley; *III* the northern Rhine graben;   *IV* basin and range province of North America

sitional transformation from the basement of the Yangtze Paraplatform (formed of the Banxi Group) in SE. Guizhou and W. Hunan into a platform cover. This was discussed by Ren Jishun in 1962 (unpublished data) The latter can be illustrated by the transitional transformation from a platform into Tianshan Geosyncline in the Hariktau area during the Sinian-Cambrian period (see Chap. V).

Tectonic migration is likewise a common phenomenon and includes geosynclinal migration, migration of the deposition centre of basins, migration of earthquake epicentres etc. The concept of tectonic migration proposed by Jiang Chunfa in 1962 (unpublished data) suggests that tectonic movements are manifested as successive regular displacements in certain directions, in certain regions and at certain stages of the development of the

Earth's crust. This is demonstrated by various geological migrations related to tectonic movements, including geosynclinal migration, migration of the folding episode, migration of the deposition centre, migration of magmatism, migration of metallogeny, migration of faulting and migration of earthquakes. This conception of tectonic migration has been developed on the basis of some migration phenomena pointed out earlier by other workers. What is different is that in the past migration phenomena were discussed individually from the angle of one's own speciality, without taking into account all kinds of migration and their interrelations.

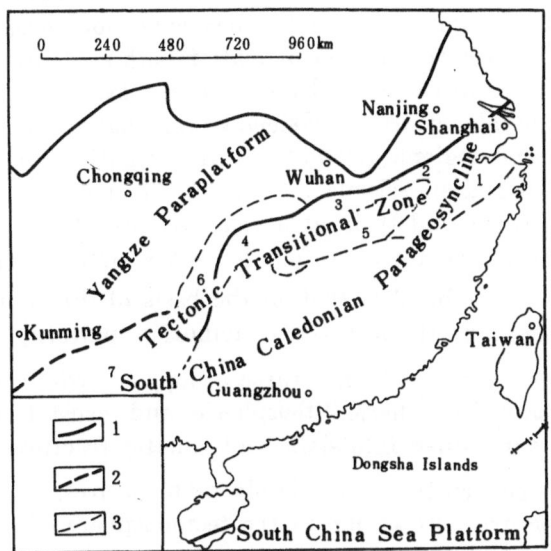

Fig. 16. Tectonic transition between the Yangtze Paraplatform and the South China
Caledonian Geosyncline. (After Ren Jishun 1964)

1 boundary between first-order tectonic units;
2 inferred boundary between first-order tectonic units;
3 boundary between second-order tectonic units.

1 Western Zhejiang; *2* Southern Anhui; *3* Xiushui River Valley; *4* Zhishui River Valley; *5* Jiuling uplift; *6* Xuefeng uplift; *7* Youjiang area

A preliminary analysis indicates that the tectonic migration of the gigantic geosynclinal belts in China shows two most important regularities: (1) Transversely, the migration is asymmetrical, with the axis of migration moving southward; in other words, the southward migration of the geosyncline between two platforms was more prominent and important than its northward migration. (2) Longitudinally, the migration occurs mainly from west to east, as in the case of the Central Asian-Mongolian Geosyncline in the Paleozoic, but from east to west as in the case of the Yunnan-Tibet Geosyncline in the Meso-Cenozoic.

It can be seen from above that tectonic migration is universal for crustal movements. Tectonic migration not only shows the orientation of the motion of surface matter, but, more important, it also reflects the process of the motion of matter at the depth of the Earth's crust and the mantle.

I.2.7 Deep Fractures

Peive (1956) systematically discussed the characteristics, classification and implications of deep fractures. In the early 1960's, Zhang Wenyou (1960), Huang Jiqing (1960) and others successively published a series of papers introducing the conception of deep fractures into the realm of geotectonic studies of China. The presence of deep fractures is an indisputable objective fact despite its denial by some geologists in subsequent years. Moreover, deep fractures have attracted more and more attention because of their long duration of development, long-distance extension, great depth of impact, close relation with the motion of matter both at depth and at the surface, great tectonic significance, and their control over the formation and distribution of mineral resources. Extensive regional geological surveys, mineral exploration and regional geophysical investigation, particularly aeromagnetic survey and deep seismic sounding have not only proved the presence of deep fractures and their great significance, but also provided abundant basic information for a profounder study of deep fractures.

Deep fractures may be classified on the basis of different factors: location, depth, age, mechanical character, distribution pattern, etc.

Deep fractures are divisible into three groups according to the depths of their occurrence: translithospheric, lithospheric and crustal fractures. The last group can be subdivided into sialic and simatic fractures.

Translithospheric fractures are simply plate sutures. They are the largest first-order fractures occurring in the tectonosphere of the Earth. They cut through the lithosphere and penetrate deep into the asthenosphere, generally constituting boundaries between oceans and active continental margins. Fractures of this group still active at present are generally distributed along deep-sea trenches and are associated with volcanic chains and deep-focus earthquakes. The Benioff Zones round the Pacific Ocean are currently active translithospheric fractures with a maximum depth of over 700 km. Deep fracture belts of this group formed in the course of evolution of the Earth are often found to contain well-developed ophiolite suites and are associated with high-pressure, low-temperature and high-temperature, low-pressure paired metamorphic and mélange belts. For instance, the Yarlung Zangbo River Deep Fracture Belt in Tibet is a typical translithospheric fracture belt.

Lithospheric fractures are also of great magnitude. They penetrate into the top part of the upper mantle or cut through the lithosphere, but do not distinctly enter the asthenosphere. Ultrabasic rocks are often found along these fractures, but ophiolitic suites are very rare.

Simatic fractures cut into the simatic layer or pass through the whole Earth's crust, but they do not enter the upper mantle. Along these fractures basaltic flows are common but ultrabasic rocks are very rare. The Cangdong Fracture in Hebei Province, for example, has been proved by deep seismic sounding to be a simatic fracture.

Sialic fractures cut through the sialic layer but do not distingctly enter the simatic layer. Granitic rocks formed by the re-melting and intrusion of crustal material are distributed along these fractures. Typical of such fractures are those of the fracture system along the southeastern coast of China.

Judging from their mechanical nature, deep fractures can be divided into three types: tensional, compressional and shearing. Fractures of a transitional nature, such as compression-shear, tension-shear, shear-tensional and shear-compressional, are assigned respectively to one of these three types in accordance with what specific mechanical character inherent in them is predominant. Fractures of continental rift valleys and oceanic rift valleys are tensional fractures. Deep fractures of the West Pacific island arcs are of predominantly compressional nature. The Altun Fracture, the Tancheng-Lujiang Fracture, the San Andres Fracture and transform faults of oceanic mid-ridges, however, are typical of shear fractures.

Deep fractures may occur in different geological times or tectonic cycles. It is sometimes rather difficult to tell the exact age of a specific deep fracture, due to the complexity of deep-fracturing activities. In the light of the characteristics of the tectonic development of China and its current studies, the authors have temporarily distinguished three main periods of activities of deep fractures: Proterozoic, Paleozoic and Meso-Cenozoic. It is hoped that further studies in the future will make specific analysis possible of the time of formation, stage of main activities and termination period of individual deep fracture belts, as well as of their depths and characters in various geological periods. Only then shall we be able to obtain more profound knowledge of any specific deep fracture and its tectonic implications.

Deep fractures may be divided into linear and arcuate ones in accordance with their forms of extension. In general, arcuate deep fractures occur in geosynclinal fold belts, whereas linear ones are mostly found in regions of continental and oceanic platforms. The former are expressions of the flexible-deformation characteristics of tectonic mobile belts, while the latter are expressions of the rigid-deformation characteristics of stable regions. Therefore, deep fractures seen both on continents and in oceans belong to fracture networks formed of linear fractures, and those seen in areas between mobile continental margins and their adjoining oceans and in the Cenozoic fold belts are belts of arcuate fractures. All this is clearly demonstrated on geological and geomorphological maps as well as on Landsat photographs.

Deep fractures can be distinguished as either exposed or concealed. Traces of the former can be identified on the surface, while the latter can only be inferred by indirect approaches by geological and geophysical methods as well as by interpretation of Landsat photographs. The abundant data obtained from deep geophysical surveys conducted in recent years have proved that concealed deep fractures separating the various layers of the Earth's crust and the upper mantle are also present in great numbers. Considerable amounts of data also indicate that any fracture belt, either in plane or in section, does not represent merely a single fault, but comprises a set or a group of fractures, ranging from a few to scores of kilometres in width.

Deep fractures have commonly undergone a process of prolonged polycyclic development, as repeatedly stressed by Huang Jiqing (1960) and Huang Jiqing et al. (1962, 1974, 1977). Now, with the accumulation of more and more valuable information and with the deepening of the various researches, we have gained an even clearer understanding. It is beyond doubt that the depth, nature and distribution pattern of deep fractures of various types are not immutable but have been steadily developing and changing in the history of the evolution of the Earth. In the course of this evolution some fractures have withered away and others have come into being; some tensional fractures have changed into compressional or shear fractures, and some compressional fractures have been transformed into tensional or shear ones, and vice versa. With the increasing depth of fracturing, crustal fractures may develop into lithospheric or even translithospheric ones, and the lithospheric fractures may become crustal or even ordinary ones as a result of shallowing of fracturing. Moreover, many facts have shown that even for one and the same fracture belt, the individual segments might have been formed at different times and may vary in character, type and depth.

Due to constant change in the total tectonic stress field of the Earth in each tectonic cycle, particularly in each mega-cycle, different tectonic cycles or stages of tectonic development have their own characteristic tectonic style of fracture. The tectonic framework of different tectonic cycles, including the distribution pattern of continents and oceans, as well as that of mobile belts and stable regions, are essentially controlled by the framework of deep fractures of different tectonic cycles. As such frameworks are often of a global nature, different geotectonic provinces are formed (here referred to as tectonic domains).

In one and the same tectonic domain, different types of deep fractures varying in depth and size show an intimate order-subordination. Accordingly, the authors classify deep fractures into four orders: fracture mega-system, fracture system, fracture zone, and individual fractures. Assigned to the first order is the fracture mega-system which controls the origin and development of tectonic domains. A fracture mega-system may comprise several fracture systems each of which in turn embodies a certain amount of fracture zones. A fracture zone again embraces a number of individual fractures. Very often in a fracture mega-system, one or several fracture systems play a leading role in the development of a tectonic domain. Such fracture systems are called principal fracture systems. The Marginal-Pacific Fracture Mega-system in eastern China, for instance, is a complex fracture system with the West Pacific Benioff Zone as its principal fracture system. Likewise, a fracture system usually has one or several principal fracture zones, and there exist one or a few principal fractures in a fracture belt.

Here special attention will be paid to the discussion of the problem of arcuate and linear fractures. Some workers (Sonder, 1956; Zhang Wenyou,

1960) emphasize the importance of linear structures, suggesting that the linear X-shaped fractures are the primary deep-seated structures, and that other tectonic types have all resulted from the development of X-shaped linear fracture networks. Based on this supposition, Zhang Wenyou has developed the theory of fault-block tectonics (Zhang Wenyou et al. 1978). However, the present authors are of the opinion that it is certainly correct to stress the importance of linear structures, particularly in rigid regions where networks of fractures of different depth and magnitude often play an outstanding or even decisive role in tectonic development, but the other aspect of the problem, i.e. the importance of arcuate structures which sometimes may be more important than the linear ones, should not be neglected.

Generally speaking, linear structures are developed in stable or less mobile regions (as in most parts of continents and oceans) and on continental margins of the Atlantic type, whereas arcuate structures tend to occur in mobile belts, especially in the active continental margins of the Pacific type and the Cenozoic fold belts of the Mediterranean. Judging from their development, linear structures occur n the stage of tensional fracturing in the early stage of the development of oceans, and arcuate structures predominate in the late stage of development of oceans, i.e. the stage of oceanic subduction or continental overthrusting as well as in the stage of continental collision. In terms of their depth, the arcuate structures seem to penetrate much deeper than the linear ones. In general, translithospheric fractures represent arcuate structures; crustal fractures are largely linear structures; and lithospheric fractures may be both arcuate and linear. The arcuate Benioff Zones around the Pacific Ocean are generally several hundreds of kilometres deep and may even reach a maximum depth of over 700 km, but the linear fractures have a far shallower depth. It is even more evident that linear fracture networks adjacent to active continental margins have resulted from the interraction (compression and shearing) between continents and oceans along the Benioff Zones. As the Atlantic-type and Pacific-type margins of oceans and continents may be mutually transformable (Mitchell and Reading 1969), the tectonic frameworks of all ages have been subject to constant evolution and changes, and therefore it is probable that the arcuate fractures on the Pacific margin and in the Mediterranean tectonic system have developed on the basis of linear structures. However, once these linear structures had developed into subduction zones, they would show a qualitative change both in depth and nature, and then they are no longer linear structures but arcuate ones. As was stated by Helde (1977), the arc-trench system (subduction zone) in the Marianas and the Philippines had actually been transformed from several N-S-trending transform faults. Conversely, the arcuate structures may also transform into linear structures. In both the new fold belts and the developing West Pacific geosynclinal belts arcuate structures predominate, while in both the old fold belts and within rigid platforms, the original arcuate fractures have been mostly replaced by linear fractures.

This is another problem worthy of discussion. Since the rise of the theory of plate tectonics, in order to analyse paleo-plate tectonics on continents, the study of mélanges, paired metamorphic belts and ophiolitic suites, which are regarded as important indications of plate sutures, has attracted great attention from many geologists. The authors are of the opinion that in the case of poor development or absence of ophiolitic suites, the occurrence of mélanges or a paired metamorphic belt alone does not necessarily imply a huge plate suture. For instance, both the Yarlung Zangbo River Deep Fracture Belt and the Bangong Lake-Nujiang River Deep Fracture Belt in Tibet are huge deep-fracture belts, but they are of different orders. The former is characterized by well-developed ophiolitic suites and represents a gigantic suture zone, while the latter, with a very poorly developed ophiolitic suite, is a suture zone of the second order. The authors believe that with the continuous deepening of scientific investigations, more and more paired metamorphic belts and mélange belts will be discovered along a number of deep-fracture belts, but they do not necessarily all represent suture zones between huge plates. Does the occurrence of ophiolitic suites definitely mean the presence of gigantic plate suture? This is not likely in some individual cases, such as the northern Qilian Geosyncline, for example. Although a fairly well-developed ophiolitic suite has been found and it is associated with a belt of glaucophane-schist, it does not appear to be a gigantic suture but a suture of second order along small plates that came into being after the split and subsidence of the Chinese Protoplatform. It is quite different from such gigantic translithospheric fractures as the West Pacific Benioff Zone. Only on the basis of a comprehensive and historical analysis and comparative tectonic study can relatively reasonable conclusions be reached to conform with the actual situation.

I.2.8 On the Problem of Tectonic Stress Fields

This is a complicated problem. With many unknown factors, its study is still in an exploratory stage at present. The present authors would like to express the following points of view:

(1) Tectonic movements are not merely an expression of the motion of matter of the surface layer of the Earth, but are reflections of the motion of matter of the mantle.

. Tectonic stress is by no means a simple mechanical force, and tectonic deformation is not a simple mechanical deformation, but involves a complicated physicochemical process. With the increasingly close combination of geological investigation day by day, with mathematics, physics and chemistry, it will become possible to gradually resolve the mysteries that confront us.

(2) Tectonic stress fields have been constantly developing and changing along with the advance of geological history. In other words, different stages of tectonic development have different tectonic stress fields. In China, for instance, the Paleozoic tectonic stress field is totally different from either the Meso-Cenozoic or the pre-Paleozoic ones. Hence, there exists absolutely no such stress field that is immutable for ever.

(3)   In each of the stages of geological history, the global tectonic stress field is unified, but it varies from region to region as well as in orders.   The interconnection of the modern global oceanic rift valleys and the regular distribution of the Cenozoic fold systems suggest an obviously unified mechanism of the global tectonic stress field.   In different regions, however, the tectonic stresses in different tectonic belts differ from one another in the manner and process of their operation, and show a certain regionalization.   This can be exemplified by the Marginal-Pacific Tectonic Domain and the Tethys-Himalayan Tectonic Domain in China, which obviously belong to two different regions of tectonic stress fields and are, in turn, divisible into stress field regions or belts of different orders.

(4)   The authors hold that the following viewpoint is open to consideration: "The major features of the surface of the Earth are produced by the interactions of processes derived by two energy sources.   The internal heat within the Earth causes motions in the mantle which influence the distribution and elevation of the continental masses. The external solar radiation drives the circulation of the atmosphere and hydrosphere, which produces the detailed sculpture of the solid Earth surface" (Wyllie 1971).   This means that there are two types of motion of matter. One is the motion of matter of the mantle (possibly the predominant one) caused by heat energy (probably originating chiefly from thermo-nuclear reaction coupled with gravitational

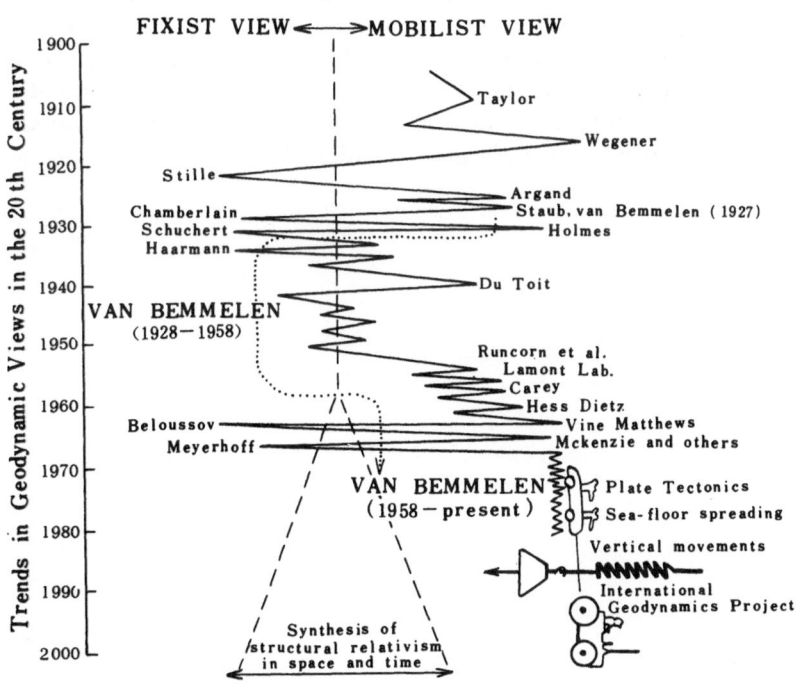

Fig.17. Trends in geodynamic views in the 20th century.

(After Van Bemmelen 1977)

action). The other is the motion of matter of the surface layer of the Earth caused by solar radiation. The combined action of the two and their interreaction and mutual transformation have resulted in the various tectonic features on the surface of the Earth. At the same time, we have to consider the motion of the Earth as a member of the solar system in the galaxy, as well as the rule of motion of the galaxy as a member of the boundless universe. For instance, one revolution of the solar system in the galaxy takes approximately 200 million years (Dai Wensai 1977), and each tectonic cycle of the Phanerozoic Eon (the Xingkai Cycle, the Caledonian Cycle, the Variscan Cycle and the Alpine Cycle) has lasted for almost the same length of time. Obviously, this is not a coincidence. There must have been some internal relations between them.

Finally, one more point is worthy of mention: whether vertical or horizontal movement is predominant in crustal movements is one of the longstanding controversies among geologists. Since the 20th century, many tectonians from West Europe and North America stress horizontal movements, but the supporters of vertical movement are still overwhelming among tectonians in the Soviet Union (Fig. 17).

The authors hold that vertical and horizontal movements reflect two different aspects of the motion of matter in the Earth's crust and mantle. Horizontal movements are an expression of the differential motion of matter in the Earth's crust and mantle in horizontal direction (including the interlayer sliding between different layers (Zhang Wenyou et al. 1978), and the extension, compression and shearing between different blocks), whereas vertical movements are an expression of the different motion of matter in vertical direction. Which one of the two is predominent depends on the difference in time, place and circumstances. Moreover, vertical and horizontal movements may be mutually transformable under certain specific conditions.

Chapter II

Subdivision of the Tectonic Cycles of China

Based on the data available, the tectonic cycles of China since the Archeozoic are subdivided as follows:

II.1 THE FUPING CYCLE

The Fuping Group is the uppermost sequence in the Archeozoic stratigraphy of China. The unconformity between the Fuping Group and Wutai Group is also considered to be that occurring between the Archeozoic and Proterozoic, which has long been accepted by Chinese geologists. Therefore, on the *Tectonic Map of China,* the Fuping Cycle is taken for the tectonic cycle of the terminal Archeozoic in China, and the Fuping Movement is represented by an angular unconformity between the Wutai Group and Fuping Group for its principal folding episode. Isotopic datings vary from 2550 to 2350 my (Cheng Yuqi et al. 1973). Its type locality is situated in the Wutai Mt. area (Plate I). Some people call this tectonic cycle the Sanggan Cycle, but since the Sanggan Group is not directly overlain by the Lower Proterozoic, the age of the "Sanggan Cycle" is undetermined.

The Fuping Group is composed of a variety of gneisses, granulites and plagioclase-amphibolites, intercalated with marbles, and is characterized by intensive migmatization, giving rise to all sorts of migmatites and migmatic granites. Metamorphic complexes assigned to the Fuping Cycle on the Tectonic Map are: the Fuping Group in the Wutai Mt. and the Taihang Mt. area, the Taishan Group in the Taishan Mt. area in western Shandong (2586 my), the Dengfeng Group in western Henan (2562 my), the Anshan Group in eastern Liaoning (2400—2800 my), and also the Wulashan Group, Sanggan Group, Qianxi Group (3400—3600 my), Dantazi Group (2400—2600 my) etc. in the Yinshan-Yanshan Mts. areas.

Data obtained in recent years demonstrated that the Archeozoic of China may be subdivided into two or more tectonic cycles. For instance, the Qianxi Group of Lower Archean, in Qianxi County of eastern Hebei, composed of metamorphics of granulite facies (originally basic and semi-pelitic rocks) gave by the whole-rock method a Rb-Sr isochron radiometric age of 3430—3670 my. This is the oldest stratigraphic sequence known in China. The upper Dantazi Group in northern Hebei, com posed of granulites, genisses and schists with marble and magnetite-quartzite intercalations (originally pelitic and carbonate rocks and basic volcanics) gave a Rb-Sr isochronic age of 2400—2600 my.

The Fuping Cycle was an important geological time in the formation of the continental crust of China, and it is most likely that the age of the lower Qianxi Group reflects the initial formation period of the Chinese mainland. The Fuping Cycle was also the most important epoch for iron-ore formation, and the well-known Anshan-type iron ores were formed in this tectonic stage.

II.2 THE WUTAI CYCLE

The geological time of the Wutai Cycle is dated at 2500—2000 my. The main folding episode is the Wutai Movement, represented by an unconformity between the Wutai Group and Hutuo Group in the Wutai Mt. area, Shanxi Province, with an isotopic age of approximately 2000 my (Cheng Yuqi et al. 1973).

The Wutai Group consists of granulites, amphibolites, amphibole-schists and green schists. The metamorphic grade is predominantly of amphibolite facies, partly of green-schist facies. The Wutai Group is unconformably underlain by the Fuping Group and is again unconformably overlain by the Hutuo Group.

Metamorphic complexes assigned to the Wutai Cycle in *the Tectonic Map of China* are: the Erdaowa Group in the southern part of the Inner Mongolia Autonomous Region (the isotopic dating on zircon from the basal conglomerate gives an age of 2350 my; unpublished data, Shen Qihan); the Kuandian Group in eastern Liaoning (the upper age limit is dated at 2040 my; unpublished data, Lin Weixing); the Lüliang Group in the Lüliang Mt. area, the Jiangxian Group in the Zhongtiao Mt. area and the Zhuzhangzi Group in the Yanshan Mt. area. The Jiaodong Group in eastern Shandong was formerly correlated with the Wutai Group and we will follow suit here, but precise data of isotopic determinations are lacking.

The Wutai Cycle was very important in the formation of the continental crust of China. Subsequent to the Wutai Cycle, most areas of the Sino-Korean Platform were practically consolidated. The Wutai Cycle represents another metallogenetic epoch for the formation of the Anshan-type iron ore deposits in our country.

II.3 THE ZHONGTIAO CYCLE

The geological time for the Zhongtiao Cycle is 2000—1700 my, the main folding episode being the Zhongtiao Movement (Wang Zhi et al. 1957) represented by an unconformity between the Zhongtiao Group (corresponding to the Hutuo Group) in the area of the Zhongtiao Mountain and the Xiyanghe Group in southern Shanxi. The isotopic age is approximately 1700 my.

In the past, the tectonic movement in the northern part of China between the Early and Late Proterozoic was formerly termed the Lüliang Movement (Lee 1939), its corresponding tectonic cycle the Lüliang Cycle, with its type locality in the Lüliang Mt. Later, it was found that the Lüliang Mt. area was short of Late Proterozoic strata, the Cambrian sequences overlying directly and unconformably the Archean or the Lower Proterozoic. Thus, at the First All-China Congress on Tectonics held in 1965, Yin Zanxun et al. (1965) suggested that the term Lüliang Movement should no longer be used; but in the area of the Zhongtiao Mountain, it is unanimously acknowledged that the unconformable relationship between the Zhongtiao Group and its overlying strata (i.e. between the Upper Proterozoic and the Lower Proterozoic) is pronounced, based on the observations made by Wang Zhi (1957), Ma Xingyuan (1957), Zhang Bosheng (1958), as well as by workers from the Regional Geological Mapping Party of Shanxi Province. Therefore, since 1976, we have adopted the term "Zhongtiao Cycle" (Huang Jiqing et al. 1977) to replace "Lüliang Cycle."

Over a long period of time, different opinions existed in China in connection with the age assignment of the Hutuo Group and its equivalents, and with their relationship with the Sinian stratigraphy of Jixian County. According to the studies by the Laboratory of Isotope Geology of the Guiyang Institute of Geochemistry, Academia Sinica (1977), the radiometric age for the basal part of the Changcheng System, Jixian County, is dated at 1950 ±50 my, and that for the Dahongyu Formation from the Nankou System at approximately 1700 my. Likewise, the age for subjection to intensive metamorphism in the Hutuo Group is also dated back approximately to 1700 my, which coincided with the time of the crustal movement prior to the Dahongyu Formation in the Yanshan Mt. area. Consequently, we adopted for the time being the opinion of correlating the Hutuo Group with the Changcheng System in the Jixian County area.

Based on the data available, all the folding periods of the Hutuo Group, Zhongtiao Group, Liaohe Group (in eastern Liaoning), and Fenzishan Group (in eastern Shandong) are assigned to the Zhongtiao Cycle in the *Tectonic Map of China*.

The Bayan-Obo Group (in the Inner Mongolia Autonomous Region) is also temporarily designated to the Zhongtiao Cycle, but some indications (isotopic ages ranging from 1350—1650 my, yielding stromatolites corresponding to the Gaoyuzhuang Formation and the Wumishan Formation of the Jixian System) (Liang Yuzuo, pers. commun.) show that its age may be somewhat younger, or it might probably be ascribed to the Wuling Cycle (see below).

II.4 THE WULING CYCLE

The Wuling Cycle was a newly established tectonic cycle in the compilation of the *Tectonic Map of China* on a scale of 1:4,000,000. Its main folding episode was the tectonic movement (Wuling Movement) as represented by an unconformity between the Lengjiaxi Group and Banxi Group, both located in the Wuling Mt. area, northwestern Hunan. This movement was also named the Dongan Movement by Wang Henian (Wang Henian 1961), but considering that the term "Wuling Movement" originated earlier (Yin Zanxun et al. 1965) than "Dongan Movement", we still call this tectonic cycle Wuling Cycle, to which the Fanjing Movement between the Fanjingshan Group and Banxi Group in eastern Guizhou, and the tectonic movement between the Upper and Lower Shennongjia Groups in western Hubei roughly correspond. At present, data of isotopic determinations for the movement in the area of northwestern Hunan are lacking. According to the age of 1332 my obtained by U-Pb isochron techniques, the upper age limit of the Wuling Cycle is temporarily determined at 1400 my.

It is generally believed that the Wuling Movement corresponds roughly to the Sibao Movement, however, the recent isotopic determinations for the Bendong pluton, which intruded the Sibao Group and then is unconformably overlain by the Banxi Group, gave an age of 1063 my, based on the related information unpublished by Yichang Institute of Geology and Mineral Resources, Chinese Academy of Geological Sciences, illustrating that the age for the Sibao Movement ranges approximately from 1000 my to 1100 my (Whether this is the same age as for the Wuling Movement needs further investigation).

II.5 THE YANGTZE CYCLE

As a tectonic cycle at the end of the Proterozoic with its type locality situated in eastern Yunnan, the Yangtze Cycle consists of two important orogenic phases, i.e. the Jinning Movement and the Chengjiang Movement. The former is referred to as the tectonic movement represented by an angular unconformity between the Kunyang Group and the Chengjiang Sandstone with an isotopic age of approximately 850 my, while the latter is referred to the tectonic movement as represented by an unconformity between the Chengjiang Sandstone and the Nantuo Tillite, its isotopic age being about 700 my.

After the Jinning Movement, a greater part of the Yangtze Paraplatform was consolidated and, after the Changjiang Movement the whole paraplatform was consolidated. The two movements constitute one tectonic cycle, the Yangtze Cycle (Huang Jiqing et al. 1974).

In addition to the Yangtze Paraplatform, those ascribed to the Yangtze Cycle in the *Tectonic Map of China* include: the Wudang, Kuanping and Taowan Groups in the Qinling Mts, the Huangyuan Group in the central Qilian Mts, the Kawabulak Group in the central Tianshan Mts, the Aierjigan Group in the Tarim Basin, etc.

The Yangtze Cycle has been verified as a tectonic cycle of epoch-making significance in the history of the evolution of the tectonics of China (Huang Jiqing et al. 1977). It finally gave rise to the formation of the Chinese Protoplatform (see the ensuing chapters), on which the most important areas of phosphorite deposits (Sinian to Cambrian) occur.

II.6 THE XINGKAI CYCLE

The Xingkai Cycle was a newly established tectonic cycle in the compilation of the *Geotectonic Map of China* on a scale of 1:10,000,000 (1976). Its time range varies from the Sinian, during which tillites began to form, to the early part of the Middle Cambrian, with its folding episode taking place in between the Early Cambrian and Middle Cambrian as represented by the type locality in the southeast of the Xingkai Lake, where the Sinian—Early Cambrian miogeosynclinal deposits composed of terrigenous and carbonate rocks are unconformably overlain by molasse of Middle Cambrian age (Okuneva and Renena 1973).

The Caledonian Cycle is generally regarded as the first tectonic cycle for the Paleozoic Era. Considering the fact that after the Jinning and Chengjiang Movements (800 my—700 my), either in China or in any other parts of the world (mainly in the Northern Hemisphere), the structure of the Earth's crust entered a new stage of development when the Sinian System actually had already become the first system for the Phanerozoic. Consequently, we are inclined to take the Xingkai Cycle for the first tectonic cycle of the Paleozoic.

The Xingkai Cycle was a very important tectonic cycle in the Central-Asian-Mongolian geosynclinal system. The Jiamusi Uplift and the Erguna Fold System were both the outcome of this tectonic cycle. In the western segment of the Central-Asian-Mongolian Geosyncline, the major folding episode of the Xingkai Cycle took place in between the Middle and Late Cambrian, which is generally called the Salair Movement (Usov 1936).

The Xingkai Movement was also significantly reflected in western Yunnan and the Indo-China Peninsula. It seemed, most probably, that the colossal Indo-China Block (including the South China Sea Platform) and the basement of the Shan State Block all resulted from the Xingkai Cycle, which may be preliminarily verified by the fact that their oldest blanket strata invariably started from the Middle—Late Cambrian. The basement of the Qiangtang Block in northern Tibet was also probably brought about by the same tectonic cycle.

II.7 THE CALEDONIAN CYCLE

The age limit for the Caledonian Cycle originally ranged from Early Cambrian to terminal Silurian. With the establishment of the Xingkai Cycle, we have assigned the Caledonian Cycle to the Middle or Late Cambrian—terminal Silurian.

In China, the most important tectonic movements of the Caledonian Cycle took place at the late stage of the Silurian, as reflected by the strong folding in geosynclinal systems represented by angular unconformities, for example, between the molasse of the Early and Middle Devonian Xueshan Group and the Silurian Hanxia Group in the Qilian Geosyncline; between the Lower Paleozoic (including the Silurian) and the Early Devonian Lianhuashan Formation (or the Middle Devonian Tiaomajian Formation) in

South China; and between the Jurumudi Formation (D_2) to the south of the western Junggar and the Mayileshan Group (S_{2-3}).

Another important tectonic movement of the Caledonian Cycle came about between the later part of the Middle Ordovician and Late Ordovician, which had a wide sphere of influence. The most intensive folding was found in the Qilian and South China geosynclinal systems, as represented by the angular unconformities between the Yaomuoshan Group (O_3) in northern Qilian and the Yingou Group (O_2); between the Yaoshuiquan Group (O_3) and the Chapu Group (O_2) in Laji Mt. (Gulang Movement); and between the Shacun Group (O_3) and O_1-O_2 in southern Jiangxi. This same tectonic movement in Burhan-Budai and in the northern Daba Mt., though not so strong, gave rise to a noticeable unconformity between the Upper Ordovician and its underlying strata.

The tectonic movements that took place between Middle and Late Ordovician in platform areas are represented by large-scale uplifts. For example, in most areas of the Sino-Korean Platform, the Middle Carboniferous Benchi Series directly overlies the Middle Majiagou Limestone; in some areas of the Yangtze Platform a disconformity between the Upper and Middle Ordovician is discernible.

Thus, in China, there were at least two major tectonic movements in the Caledonian Cycle: one occurred between the late part of the Middle Ordovician and Late Ordovician (Early Caledonian Movement), another came about between the terminal Silurian and Early Devonian (Late Caledonian Movement), this latter being more important. Thus we have subdivided the Caledonian Cycle into two subcycles: the Early Caledonian Subcycle and the Late Caledonian Subcycle. The geosynclines such as the Qilian Mts., South China area, Northern Qinling and Northern Daba Mts. were all transferred into fold belts through the late Caledonian Movement. Therefore, all Caledonian geosynclinal fold belts (systems) of China are Late Caledonian in age.

II.8 THE VARISCAN CYCLE

The Variscan Cycle refers to the tectonic cycle from Early Devonian to Permian. As this cycle resulted in the formation of primitive Asia, or Pal-Asia, hence it is the most important tectonic cycle in the Pal-Asian tectonic domain.

Based on the fact that the Tianshan Mts. is one of the typical Variscan Geosynclinal Fold Systems of China, Hu Bing (Compilation Group of Xinjiang Geological Bureau 1978), Yin Zanxun (1978) and Luo Fazuo (1981) named the Variscan Cycle the Tianshan Cycle, which, according to their opinion, involves 8 to 9 phases. The subdivision of the Variscan Cycle into four stages, i.e. the early, middle, late and terminal stages, was proposed on the Tectonic Map of China, respectively corresponding to tectonic movements at the end of the Devonian, end of Early Carboniferous, end of Late Carboniferous, and the end of Permian. It should be pointed out that in China geosynclines of the Early Variscan generally ended during the Late

Devonian, and geosynclines of the Late Variscan ended at the close of the Early Permian.

1. *The Early Variscan Cycle*

In general, its major episode took place in the Late Devonian. The Maoniushan Formation (D_3^1) of geosynclinal type on the northern margin of the Qaidam Basin is unconformably overlain by the Amunike Formation ($D_3^2 C_1$) of molasse character, forming the Early Variscan fold zone.

In the Altay Geosyncline the Early Carboniferous Hongshanzui Formation rests disconformably on geosyncline-type deposits. The Hongshanzui Formation, 500 to 700 m thick, is made up of intermediate acidic pyroclastic rocks of marine facies, but part of the mottled sandy slate in the upper part is probably non-marine. In the light of the characteristics described above, and in combination with the geological data published by the Soviet Union and the Mongolian People's Republic, it is inferred that this formation might belong to platform type. We are inclined, for the time being, to regard the Altay Geosyncline as an Early Variscan foldbelt. Though the Early Variscan had not arrested geosynclinal development in other regions, yet the influence it exerted was also quite considerable. According to Luo Fazuo (previously cited), on the northern slope of the Borohoro Mt. and the southern margin of the Ili Depression at the west end of the northern Tianshan Mt., the Early Carboniferous Dahalajunshan Formation rests unconformably on the Late Devonian Tuosikuertawu Formation. This tectonic movement is also common for the southern part of the Tianshan Mts. as represented by the Gancaohu Formation (C_1) unconformably overlying the Puochengzi Formation (D_3). The depression belt of the Silurian period, which was situated to the north of a line linking Aihui and Erlian in the Greater Hinggan Mts., had been uplifted into land due to this tectonic movement. In the north of the Qinling Mts., major tectonic movements took place between the Middle and Late Devonian as shown by the fact that the Xihanshui Group (D_2) is unconformably overlain by the Dacaotan Group (D_3).

2. *The Middle Variscan Cycle*

Its major episode took place at the end of the Early Carboniferous, and in some places it occurred within the Visean stage. This tectonic movement is quite important, it not only converted the Junggar geosynclinal sedimentation into a fold belt, but was also furnished with an extremely extensive sphere of influence, the Savur Movement in the Soviet Union and the Shizu Movement in Japan, for example. In Vietnam, the movement is represented by an unconformity between C_2 and C_1/D. All these indicate the manifestation of the Middle Variscan. In the western Junggar, the geosynclinal deposits of the Nanmingshui Formation (C_1) is unconformably overlain by the non-marine deposits of the Klagong Formation (C_2), while in the eastern Junggar, the Nanmingshui Formation is unconformably overlain by the non-marine volcanic rocks of the Batamayineishan Formation (C_2), indicating the end of the evolutional history of the Junggar Geosyncline. In the Qoltag of the Tianshan Mts., the Yamansu Formation (C_1) is unconformably overlain by the Dikaner Formation (C_2). In the northern Qinling, the geosynclinal deposits of Early Carboniferous are unconformably overlain by the coal-bearing beds of the Caoliangyi Formation (C_2).

3. *The Late Variscan Cycle*

Its major episode took place at the end of the Late Carboniferous and prior to this episode there were also found traces of tectonic movement occurring between the Middle and the Late Carboniferous in certain areas, as represented by the fact that the Shirengou Formation (C_3) is unconformably overlain by the mottled clastic rocks of the Aqikebulak Group (P_1) in the northern Tianshan, thus converting most of the area from a geosyncline into a fold belt. Due to the influence of this tectonic movement, the Hariktau area in southern Tianshan was also transformed into a fold belt. In the Inner Mongolia Autonomous Region the unconformity between the Amushan Formation (C_3) and the Zhesi Formation (P_1) represents the same movement. A similar relationship is shown near Dakendaban in the southern Qilian Mts., where clastic and volcanic rocks (P_1) unconformably overlie green schists (C_3).

4. *The Terminal Variscan Cycle*

Its major episode took place between the Early and Late Permian, when geosynclinal sedimentation in the Pal-Asian Tectonic Domain came to an end completely. One tectonic movement prior to the major episode occurred between the early and late stages of the Early Permian, while a tectonic movement after the major episode occurred at the end of the Permian, but was confined only to a limited area. The geosynclinal deposits of the Early Permian Kurgan Formation are unconformably overlain by the molasse of the Biyoulebaoguzi Group (P_2) in the southern Tianshan. In the Beishan Mts., the Jinta Formation is unconformably overlain by the Fangshankou Group. In the Greater Hinggan Mts., the Liutiaogou Formation is unconformably overlain by the Sunjiafen Formation. All this represents the major episode. Unconformities between the Early and Late Permian are also found in the Jinsha River and the Kaixin Range, Tibetan Plateau. In the western Kunlun and Qiman Tagh, tectonic movements occurred probably between the Early and Late stages of the Early Permian, leading to the conclusion of geosynclinal sedimentation.

II.9 THE ALPINE CYCLE

In 1945, Huang Jiqing divided the Alpine Cycle of China into three subcycles, that is, the Indosinian Subcycle, the Yanshanian Subcycle, and the Himalayan Subcycle. We basically adopted this subdivision in the compilation of the Tectonic Map of China, but have made some modifications in accordance with recent information.

II.9.1 The Indosinian Subcycle

Since Huang established the Indosinian Subcycle, geological surveys and monographic studies done over the past 30 years or so have increased the understanding of this tectonic cycle. The Indosinian Movement has been confirmed as an epoch-making tectonic movement, while China has proved to be one of the regions with its best development. The tremendous geosynclinal fold belts, such as the Sanjiang, the Songpan-Garze and the Qinling belong to the Indosinides, and the Post-Caledonian platform cover (D-T$_{1-2}$) in Guangxi, Hunan, Jiangxi and Guangdong was strongly affected by Indosinian tectonism.

The most important folding of the Indosinian Cycle occurred in the early stage of the Late Triassic (possibly between Carric and Noric). This is represented by the unconformity below the Shizhongshan Formation (called the Jiezha Group in Qamdo) of Late Triassic in western Yunnan (Ren Jishun et al. 1966), below the Late Triassic Anyuan Group (Genkou Group) in South China (Huang Jiqing et al. 1936—37), and below the non-marine volcanic rocks and coal measures of Late Triassic in the southern part of the Burhan Budai Mt. and Jishi Mt. Another important folding had occurred from the end of the Triassic to Early Jurassic as shown most distinctly in the area of the Longmen Mt. in western Sichuan, where the Baitianba Formation of Early Jurassic is unconformably underlain by the folded Xujiahe Formation of Late Triassic age; while in the lower Yangtze an angular unconformity exists below the Early—Middle Jurassic Xiangshan Group, with the Fanjiatang Formation being of Late Triassic age.

Earlier tectonic movement in the Indosinian Cycle came about in the Middle Triassic. According to Duan Xinhua (unpublished data), a folding movement occurred in the area near Jinghong and Mengla in western Yunnan during the Middle Triassic epoch, as shown by an angular unconformity between the Yibi Formation (Ladinic) and the Bangsha Formation (Anisic). The facts that the folding in Qinling Geosyncline probably occurred in the late part of the Middle Triassic, that the great marine regression occurred in the Middle Triassic on the Yangtze Paraplatform resulting in the lacuna of the Ladinic stage in a wide area, as well as the rapid downwarping of the Songpan-Garze, Yujiang and other Indosinian geosynclines since the Middle Triassic, indicate that there had been an important tectonic movement during the Middle Triassic, which might possibly be the earliest major tectonic movement for the Indosinian Cycle.

Thus, the Indosinian Cycle at present can be subdivided into at least three major orogenic phases, namely, during the Middle Triassic, from post-Mid-Triassic to the early stage of the Late Triassic, and from the end of the Late Triassic to the Early Jurassic. The second one appears to be the most important..

The Indosinian Cycle represents an important coal-forming period in the regions of South China and Qinghai-Tibet. The well-known coal fields, such as Yongren and Yipinglang in Yunnan Province, Anyuan Coal Field in

Jiangxi Province, the Tumengela Coal Field in Tibet, etc., are products of this cycle.

II.9.2 The Yanshanian Subcycle

The Yanshanian Subcycle derives its name from the "Yanshan Movement", already very well known in Chinese literature, and is of great importance to the tectonic development and formation of mineral resources in the Mesozoic Era in China. Unfortunately, the Yanshan area is not a Mesozoic geosyncline, and Jurassic and Cretaceous marine strata are absent there as well as in eastern China, making it difficult to carry out a through investigation and study of the Yanshanian Cycle. For this reason, we have subdivided the Yanshanian Subcycle preliminarily into the early, middle and late stages, on the basis of the analysis of the Mesozoic Marginal Pacific geosynclines in eastern Asia and the Tethys geosynclines in Yunnan, and in combination with analysis on the recent advances in the study of the Meso-Cenozoic stratigraphy of China.

The Early Yanshanian Subcycle, ranging in age from the Early Jurassic to the early stage of the Middle Jurassic.

The Middle Yanshanian Subcycle, ranging in age from the late stage of the Middle Jurassic to the early stage of the Early Cretaceous.

The Late Yanshanian Subcycle, ranging in age from the late stage of the Early Cretaceous to the early stage of the Late Cretaceous.

1. *The Early Yanshanian Subcycle* (J_1—*Early* J_2)

Its major folding episode occurred from the late stage of the Middle Jurassic to the early stage of the Late Jurassic. In the Tethys Geosyncline, it refers to the tectonic movement represented by the universal overlapping unconformity between the Liuwan Formation of the late stage of the Middle Jurassic or its equivalents in the upper reaches of the Nujiang River in western Yunnan and the underlying strata, which in the Marginal Pacific Tectonic Domain refers to the tectonic movement as characterized by an unconformity between the Longzhuagou Group of the late stage of the Middle Jurassic or its equivalents and the underlying strata. It seems that both the unconformity between the Mentougou Coal Measures of the Early and Middle Jurassic and the volcanic rocks of the late stage of the Middle Jurassic west of Beijing, and the regional unconformity below the volcanic rocks of the Late Jurassic and Early Cretaceous along the coastal areas of Zhejiang, Fujian and Guangdong provinces are indications of this tectonic movement. The Early Yanshanian Subcycle was one of the major coal-forming periods in North China, the Mentougou coal field west of Beijing and the Fangzi coal field in Shandong Province, and others all being the products of this tectonic stage.

2. *The Middle Yanshanian Subcycle* (*Late* J_2-*Early* K_1)

Its major folding episode took place during the Early Cretaceous. In the Sikhote-Alin Geosyncline it refers to the Hauterivian Movement occurring in the Hauterivian stage, slightly earlier than that for the Sakawa

Movement of Japan. This movement in the Tethys geosyncline had stopped geosynclinal development in the area of Nagqu, N. Tibet. In Amdo and else-where, fossiliferous strata of the Early Cretaceous may be seen to overlie unconformably the ultrabasic rocks emplacing the Middle and Late Jurassic (probably including the early stage of Early Cretaceous) sequences. All the unconformities below the conglomerates of the Sunjiawan Formation in western Liaoning Province, below the Shaxian Formation in Fujian Province, and below the Dongjing Formation in Hunan Province are results of this movement.

The Middle Yanshanian Subcycle was a period in which the strongest magmatic activities of eastern China in the Marginal Pacific Tectonic Domain took place, and extensive volcanic belts, together with large-scale granite batholiths in the Greater Hinggan Mts. and along the southeastern coast areas of China, were formed in this period. A great deal of famous ore deposits in eastern China, such as the tungsten ore deposits in southern Jiangxi, the Dexing copper ore deposits in northern Jiangxi and the iron ore deposits in the Lower Yangtze valley are products of this period. The later part of this subcycle was one of the major coal-forming periods in North China and Northeast China. The coal fields of Fuxin, Jixi and other areas were formed in this period.

3. *The Late Yanshanian Subcycle* (*Late K_1—Early K_2*)

Its major folding episode occurred in the late stage of Late Cretaceous (Campanian-Maestrichtian). The unconformities below the Sifangtai Formation in the Songliao Basin of Northeast China and below the Nanxiong Group in the Nanxiong Basin of Guangdong Province represent this movement. The principal folding phases for the Sikhote-Alin and Gangdisê-Nyainqêntang-lha geosynclines belong to the same category. During the Late Yanshanian Subcycle, huge quantities of petroleum accumulated in the Songliao Basin (the famous Daqing Oil Field), while large salt deposits occur in Jiangxi Province.

II.9.3 The Himalayan Subcycle

Based upon data of geological surveys and scientific expeditions conducted in Sichuan, Qinghai, Yunnan and the Tibet Autonomous Region, as well as on the work carried out in areas of island arcs and marginal seas of East China in recent years, we subdivide the Himalayan Subcycle in to the early and late stages, including three phases of important tectonic movements. The Early Himalayan Subcycle came about from the late stage of the Late Cretaceous to the middle stage of the Miocene, and the Late Himalayan Subcycle occurred since the Miocene.

There are two major tectonic movements for the Early Himalayan Subcycle. The first occurred at the end of Middle Eocene when the Tethys Sea was closed and the Himalayan Geosyncline was folded, giving rise to a strong collision between India and Asia. In North China, this movement is shown by a slight unconformity between the Shahejie and Kongdian Formations;

and in northern Jiangsu, between the Dainan and Funing Formations. The second occurred fro mthe end of the Oligocene to the Middle Miocene in the form of intensive folding, metamorphism and magmatic activities in the Himalayas; in the island-arc area, in the form of the Takachiho Movement of Early Miocene in the Shimanto Geosyncline of Southwest Japan. In the Central Range of Taiwan Province, it is shown by the folding of both the Tertiary and Lower Miocene sequences. It is also distinctly revealed in separate basins of East China, as indicated by the unconformities between the Guantao Formation (Miocene) and Dongying Formation (Oligocene) in North China; between the Yancheng Formation (Miocene) and the Sanduo Formation (Oligocene) in northern Jiangsu; and between the Da'an Formation and its underlying strata in the Songliao Basin.

The major episode of the Late Himalayan Subcycle, the Taiwan Movement, took place from the Pliocene to the Early Pleistocene when the Taiwan Geosyncline was folded. In the Himalayas and the Tibetan Plateau, the movement is represented by strong uplifting since the Pliocene. According to Huang Jiqing et al., this uplifting was accompanied by three explosive phases with three molasses being formed in the piedmont belts of the Himalayas, the Kunlun Mts., the Tianshan and the Qilian Mts., (Huang Jiqing et al. 1980). After this movement, extensive subsidence took place in East China and its marginal sea basins, eventually shaping the present distribution pattern of land and sea.

Tectonic movement after the Early Pleistocene should also be assigned to the Himalayan Subcycle, which will not be dealt with in this paper.

Chapter III

Brief Description of the Main Tectonic Units of China[4]

In 1945 Huang Jiqing made the first complete and systematic subdivision of the main tectonic units of China and gave a brilliant exposition thereof. Since then, through more than 30 years of geological work, geologists generally agree upon the scheme of subdivision of the main tectonic units of China except for a few areas where further studies remain to be done. This book represents a summary of arduous efforts of Chinese geologists. In this particular chapter we shall do our best to reflect this accomplishment as precisely as possible with the help of the *Tectonic Map of China*, a scale 1:4,000,000.

In the *Tectonic Map of China*, with the aim of making the terms of tectonic units concise and explicit and at the same time giving consideration to the established usage of Chinese geological workers, we basically accept the scheme of nomenclature proposed for *the Geotectonic Map of China*, a scale 1:1,000,000, published in the 1960's. Only a few necessary modifications and supplements are made in accordance with new advances in geological sciences in recent years.

III.1 SYSTEM OF NOMENCLATURE FOR TECTONIC UNITS

1. *In Geosynclinal Fold Regions*

First order: Geosynclinal fold regions, e.g. the Tianshan-Hinggan Geosynclinal Fold Region.

First suborder: Geosynclinal fold systems (for short called fold systems), e.g. the Tianshan Fold System.

Second order: Eugeosynclinal fold belts, e.g. the Northern Tianshan Eugeosynclinal Fold Belt. Miogeosynclinal fold belts, e.g. the Southern Tianshan Miogeosynclinal Fold Belt. Median uplift belts, e.g. the Tianshan

---

4 In Preparation of this chapter, the authors consulted not only some published literature but also a large quantity of unpublished data, especially the results of regional geological surveys of various provinces (regions).

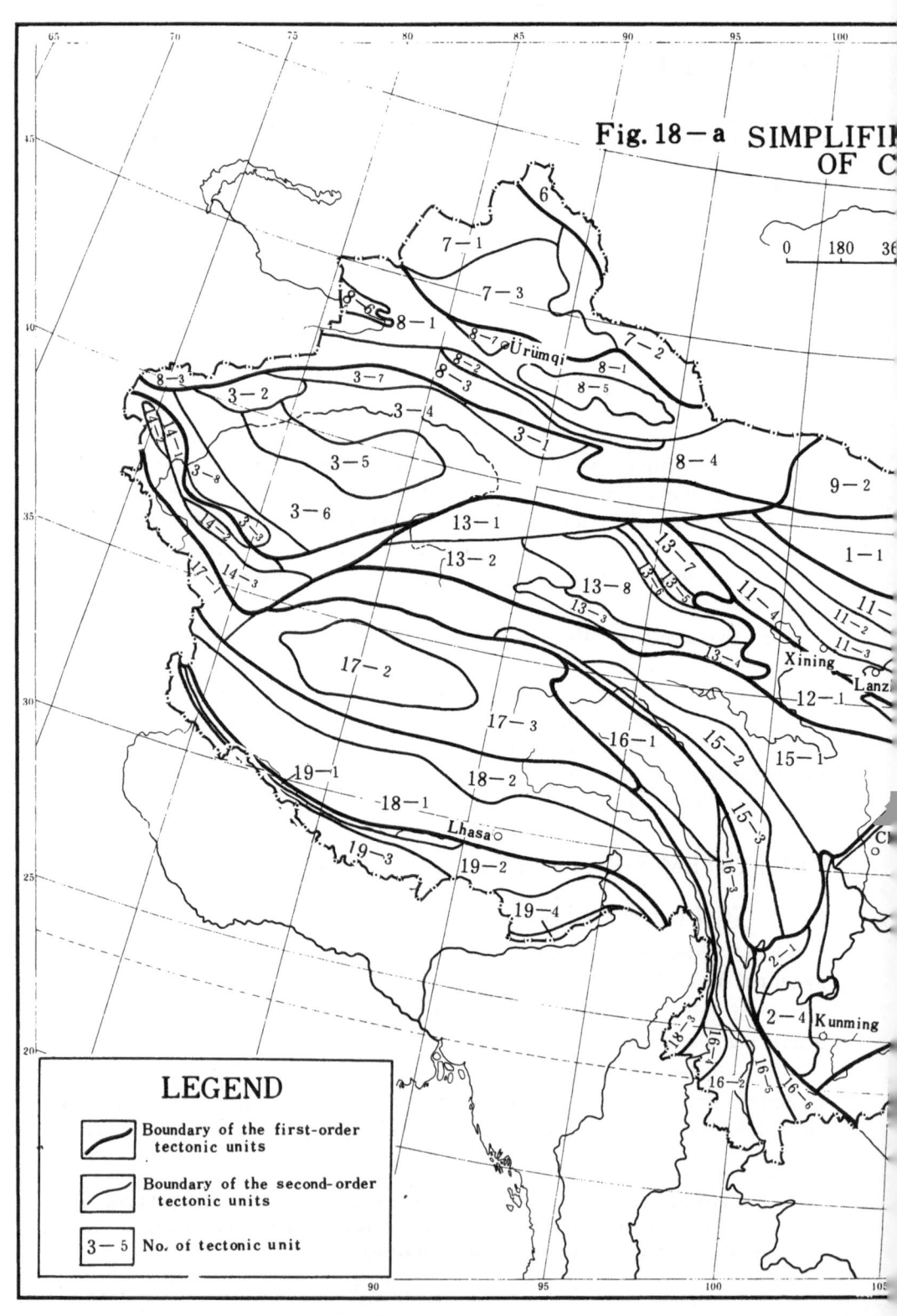

Fig. 18-a SIMPLIFI
OF C

LEGEND

Boundary of the first-order tectonic units

Boundary of the second-order tectonic units

3 — 5   No. of tectonic unit

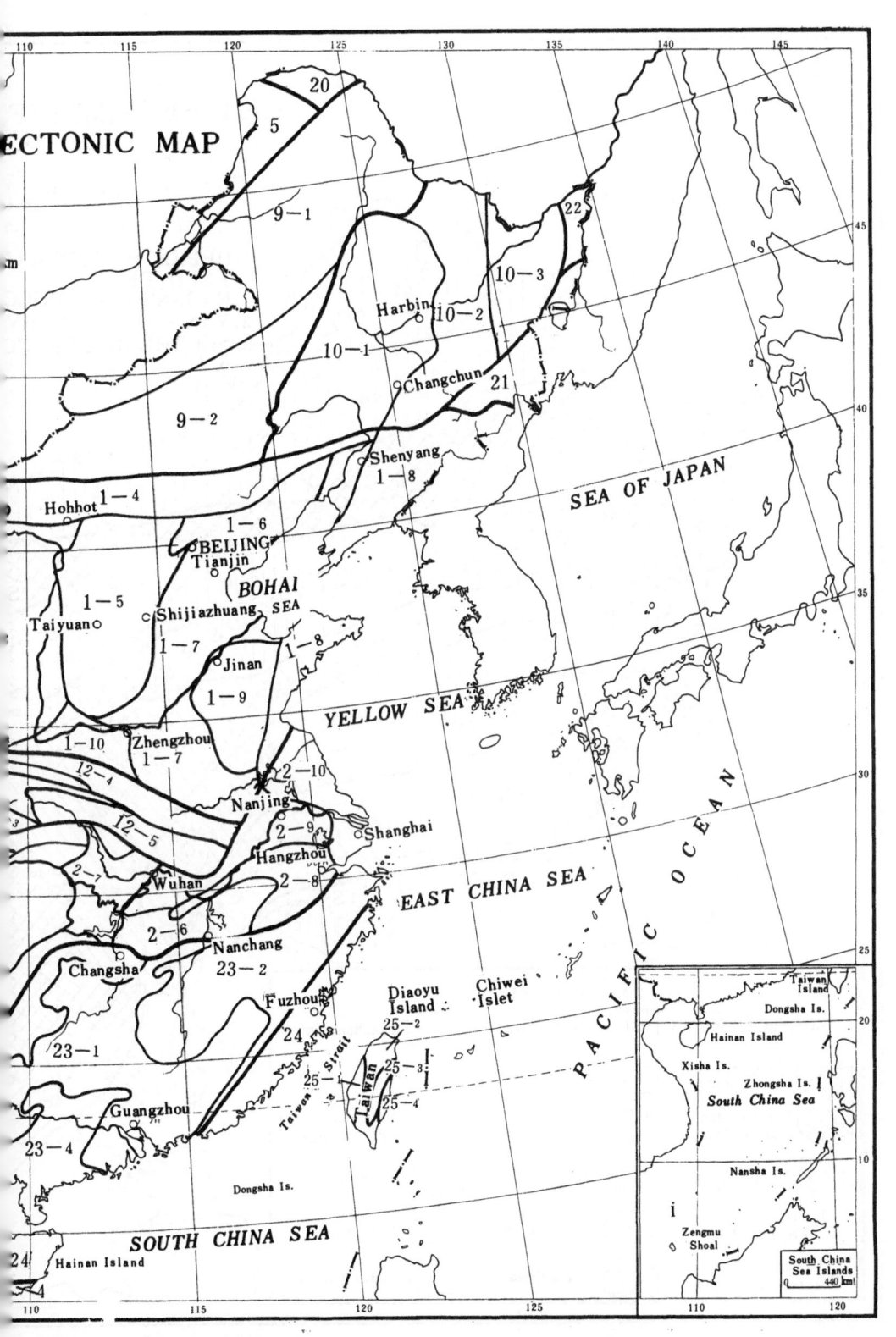

ECTONIC MAP

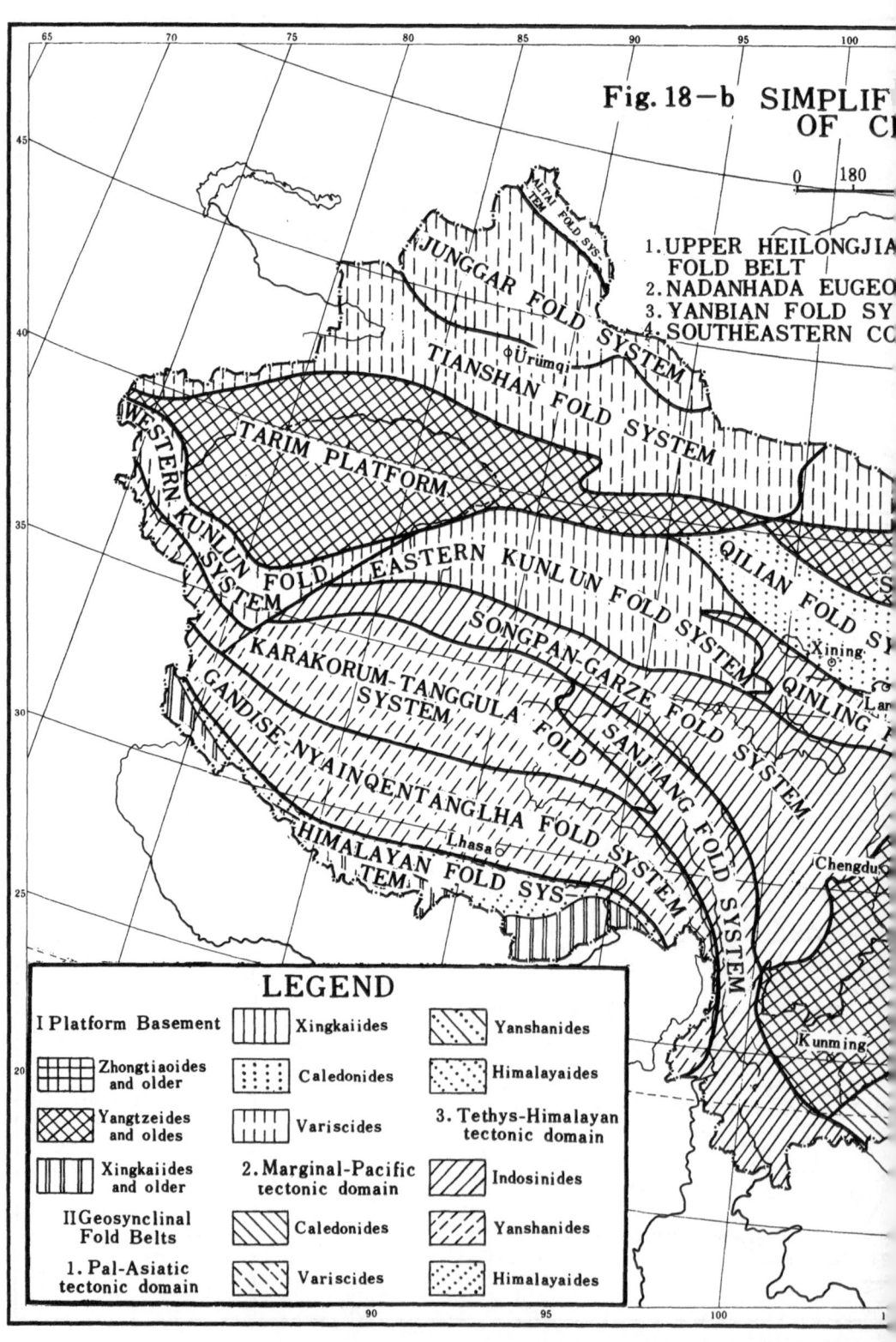

Fig. 18-b SIMPLIF
OF C

0   180

1. UPPER HEILONGJIA
   FOLD BELT
2. NADANHADA EUGEO
3. YANBIAN FOLD SY
4. SOUTHEASTERN CO

LEGEND

I Platform Basement	▭ Xingkaiides	▨ Yanshanides
▦ Zhongtiaoides and older	⁙ Caledonides	▦ Himalayaides
▧ Yangtzeides and oldes	▥ Variscides	3. Tethys-Himalayan tectonic domain
▥ Xingkaiides and older	2. Marginal-Pacific tectonic domain	▨ Indosinides
II Geosynclinal Fold Belts	▨ Caledonides	▨ Yanshanides
1. Pal-Asiatic tectonic domain	▨ Variscides	⁙ Himalayaides

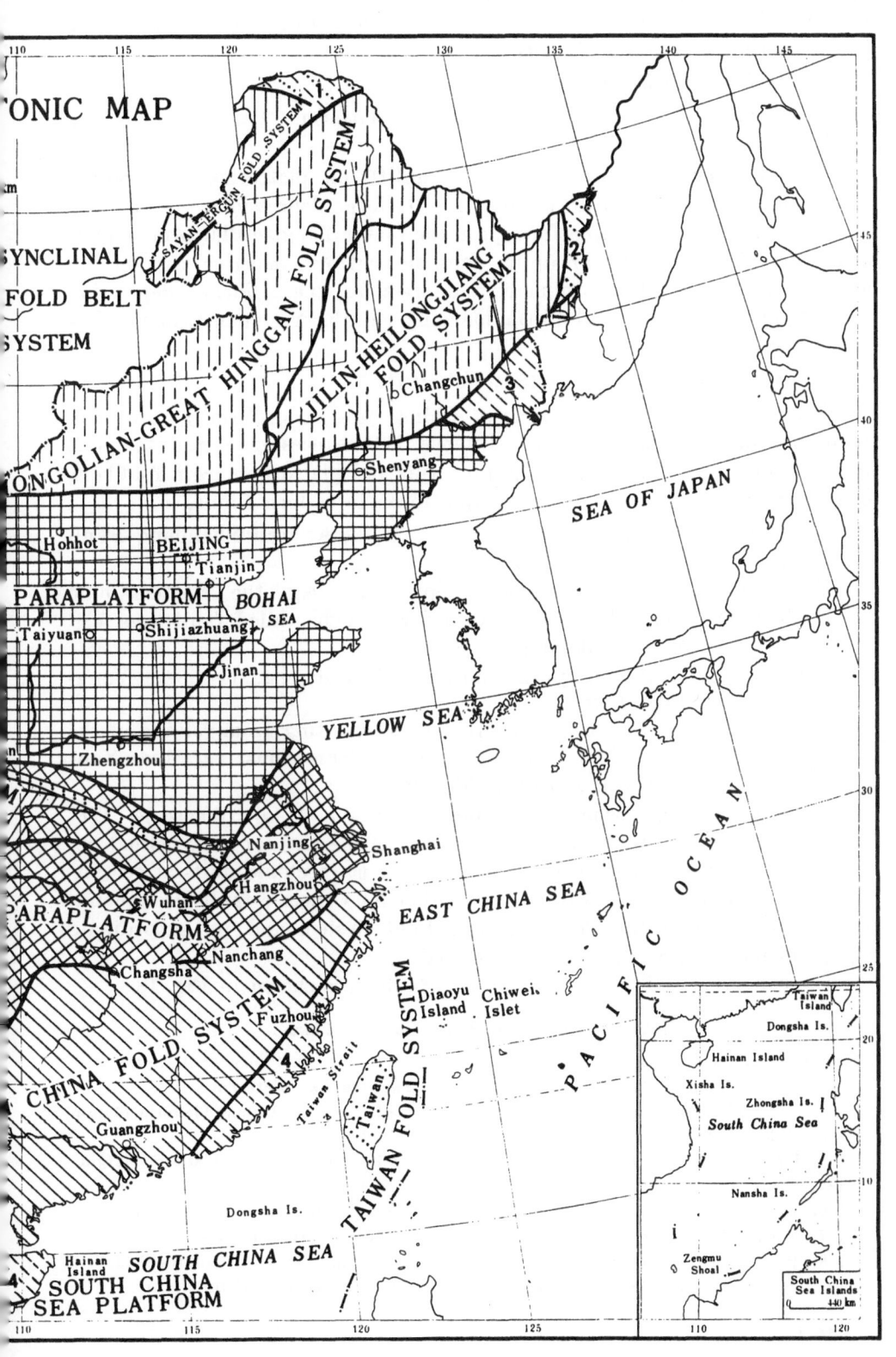

TONIC MAP

SYNCLINAL
FOLD BELT
SYSTEM

SAYAN-ERGUN FOLD SYSTEM

MONGOLIAN-GREAT HINGGAN FOLD SYSTEM

JILIN-HEILONGJIANG FOLD SYSTEM

1

2

3

Changchun

Shenyang

SEA OF JAPAN

Hohhot   BEIJING
              Tianjin

PARAPLATFORM   *BOHAI*
                              SEA

Taiyuan

Shijiazhuang

Jinan

YELLOW SEA

Zhengzhou

PARAPLATFORM

Nanjing   Shanghai

Hangzhou

Wuhan

EAST CHINA SEA

Nanchang

Changsha

Diaoyu   Chiwei
Island   Islet

CHINA FOLD SYSTEM

Fuzhou

4

TAIWAN FOLD SYSTEM

PACIFIC OCEAN

Taiwan Strait

Taiwan

Guangzhou

Dongsha Is.

Hainan   SOUTH CHINA SEA
Island
SOUTH CHINA
SEA PLATFORM

Taiwan
Island

Dongsha Is.

Hainan Island

Xisha Is.

Zhongsha Is.

South China Sea

Nansha Is.

Zengmu
Shoal

South China
Sea Islands
440 km

Median Uplift Belt. Piedmont depressions, e.g. the Urümqi Piedmont Depression. Intermontane depressions, e.g. the Turpan-Hami Intermontane Depression. Uplifts, e.g. the Jiamusi Uplift.

2. *In Platform Regions*

First order: Platforms, e.g. the Russian Platform. Paraplatforms, e.g. the Yangtze Paraplatform.

Second order: Platform uplifts, i.e., uplifts on platforms or paraplatforms, e.g. the Jiaoliao Platform Uplift. Platform depressions, i.e., depressions on platforms or paraplatforms, e.g. the Lower Yangtze Platform Depression. Axes, i.e., active linear uplift belts on platform margins, e.g. the Inner Mongolian Axis. Platform-margin depressions, i.e., linear depression belts on platform or paraplatform margins, e.g. the Yanyuan-Lijiang Platform-Margin Depression.

3. *In Continent-Margin Activization Belts*

Since the Mesozoic, most parts of China have transformed into continent-margin activization belts. Intense disturbances gave rise to fault uplifts, fault depressions, platform fold belts and platform-margin fold belts on the original, relatively stable paraplatforms in eastern China.

Fault uplifts, i.e., fault-block uplifts, e.g. the western Shandong Fault Uplift.

Fault depressions, i.e., fault-block depressions, e.g. the North China Fault Depression. Those of lower order may be called fault sags, e.g. the central Hebei Fault Sang.

Platform fold belts, i.e., platform-cover fold belts, e.g. the Lower Yangtze Platform Fold Belt.

Platform-margin fold belts, i.e., fold belts resulting from folding of the platform cover in original platform-margin depressions, e.g. the Yanyuan-Lijiang Platform-Margin Fold Belt.

For western China, what are highlighted in the 1:4,000,000-scale *Tectonic Map of China* are the tectonic features in the stage of geosynclinal development, so the system of nomenclature for the tectonic units in geosynclinal fold regions is used without exception for Cenozoic fault-block mountain systems, such as the Tianshan Mts. Fold System and its subordinate units, and no new system will be developed.

The subdivision of the tectonic units of China is shown in Fig. 18.

Table 1. Major tectonic Units of China

1. Sino-Korean Paraplatform
1-1. Alxa Platform Uplift
1-2. Ordos Platform–Margin Fold Belt
1-3. Ordos Platform Depression
1-4. Inner Mongolian Axis
1-5. Shanxi Fault Uplift
1-6. Yanshan Platform Fold Belt
1-7. North China Fault Depression
1-8. Jiaoliao Platform Uplift
1-9. Western Shandong Fault Uplift

1-10. Western Henan Fault Uplift
2. Yangtze Paraplatform
2-1. Yanyuan–Lijiang Platform-Margin Fold Belt
2-2. Longmen–Daba Platform-Margin Fold Belt
2-3. Sichuan Platform Depression
2-4. Kam-Yunnan Axis
2-5. Upper Yangtze Platform Fold Belt
2-6. Jiangnan Platform Uplift

(originally called Jiangnan Axis)

2-7. Jianghan Fault Depression

2-8. Western Zhejiang-Southern Anhui Platform Fold Belt

2-9. Lower Yangtze Platform Fold Belt

2-10. Northern Jiangsu Fault Depression

3. Tarim Platform

3-1. Kuruk Tagh Fault Uplift

3-2. Kalpin Fault Uplift

3-3. Tikanlik Fault Uplift

3-4. Northern Platform Depression

3-5. Central Platform Depression

3-6. Southern Platform Depression

3-7. Kuqa Piedmont Depression

3-8. Kunlun Piedmont Depression

4. South China Sea Platform

5. Ergun Fold System

6. Altay Fold System

7. Junggar Fold System

7-1. Western Junggar Eugeosynclinal Fold Belt

7-2. Eastern Junggar Eugeosynclinal Fold Belt

7-3. Junggar Depression

8. Tianshan Fold System

8-1. Northern Tianshan Eugeosynclinal Fold Belt

8-2. Tianshan Median Uplift Belt

8-3. Southern Tianshan Miogeosynclinal Fold Belt

8-4. Beishan Eugeosynclinal Fold Belt

8-5. Turpan-Hami Intermontane Depression

8-6. Ili Intermontane Depression

8-7. Urumqi Piedmont Depression

9. Inner Mongolian-Greater Hinggan Fold System

9-1. Greater Hinggan Eugeosynclinal Fold Belt

9-2. Inner Mongolian Eugeosynclinal Fold Belt

10. Jilin-Heilongjiang Fold System

10-1. Songliao Depression

10-2. Zhangguangzailing Eugeosynclinal Fold Belt

10-3. Jiamusi Uplift

11. Qilian Fold System

11-1. Corridor Transition Belt (Miogeosynclinal Fold Belt)

11-2. Northern Qilian Eugeosynclinal Fold Belt

11-3. Qilian Median Uplift Belt

11-4. Southern Qilian Fold Belt

12. Qinling Fold System

12-1. Southern Qinling Miogeosynclinal Fold Belt

12-2. Lixian-Zhashui Miogeosynclinal Fold Belt

12-3. Northern Daba Mt. Fold Belt

12-4. Northern Qinling Fold Belt (including Qinling Axis)

12-5. Wudang-Huaiyang Uplift

13. Eastern Kunlun Fold System

13-1. Altun Eugeosynclinal Fold Belt

13-2. Qimantagh Eugeosynclinal Fold Belt

13-3. Eastern Kunlun Median Uplift Belt

13-4. Burhan Budai Eugeosynclinal Fold Belt

13-5. Oulongbuluk Uplift Belt

13-6. Eugeosynclinal Fold Belt on the Northern Margin of Qaidam

13-7. Dakendaban Fold Belt

13-8. Qaidam Depression

14. Western Kunlun Fold System

14-1. Northern Kunlun Miogeosynclinal Fold Belt

14-2. Western Kunlun Median Uplift Belt

14-3. Southern Kunlun Fold Belt

15. Songpan-Garzê Fold System

15-1. Bayan Har Fold Belt

15-2. Yajiang Fold Belt

15-3. Yüshu-Yidun Fold Belt

16. Sanjiang (Three-River) Fold System

16-1. Wuli-Angqi Fold Belt

16-2. Lancang River Fold Belt

16-3. Jinsha River Eugeosynclinal Fold Belt

16-4 Baoshan Fold Belt

16-5. Lanping-Shimao Depression

16-6. Ailao Mt. Fold Belt

17. Karakorum-Tanggula Fold System

17-1. Linqitang Miogeosynclinal Fold Belt

17-2. Qiangtang Uplift (Qiangtang Massif)

17-3. Tanggula Miogeosynclinal Fold Belt

18. Gangdisê-Nyainqêntanglha Fold

System
18-1. Lhasa Fold Belt
18-2. Nagqu Fold Belt
18-3. Tengchong Fold Belt
19. Himalayan Fold System
19-1. Yarlung-Zangbo-River Eugeosynclinal Fold Belt
19-2. Northern Himalayan Eugeosynclinal Fold Belt
19-3. Himalayan Nappe Belt
19-4. Siwalik Piedmont Depression
20. Upper Heilongjiang Miogeosynclinal
Fold Belt (belonging to Mongolia-
Okhotsk Fold System)
21. Yanbian Fold System
22. Nadanhada Eugeosynclinal Fold Belt

(belonging to the Sikhote-Alin Fold
System)
23. South China Fold System
23-1. Jiangxi-Hunan-Guangxi-Guang-
dong Fold Belt
23-2. Cathaysian Fold Belt
23-3. Youjiang Fold Belt
23-4. Yunkai Fold Belt
24. Southeastern Coastal Fold System
25. Taiwan Fold System
25-1. Taiwan-Penghu Depression
25-2. Western Taiwan Miogeosynclinal Fold Belt
25-3. Danan'ao Median Uplift Belt
25-4. Eastern Taiwan Eugeosynclinal
Fold Belt

II.2 THE SINO-KOREAN PARAPLATFORM

The Sino-Korean Paraplatform, triangular in form and with deep fractures separating it from its neighbouring tectonic units, covers the whole territory of North China north of the Qinling Mts., the southern part of Northeast China, the Bohai Gulf, the northern part of the Huanghai Sea and
the northern part of Korea. It is bounded on the north by the Tianshan-
Hinggan Geosynclinal Fold Region ;along the deep fractures on the northern
margins of Alxa and the Inner Mongolian Axis, on the south by the Qinling
Fold System along the deep fracture on the northern margin of the Qinling
Axis and the Queshan-Feidong Deep Fracture, and on the east by the Yangtze Paraplatform along the Tancheng-Lujiang and Jiashan-Xiangshui Deep
Fractures.

The Sino-Korean Paraplatform is the oldest in China. The first continental nucleus might originate 3000 my ago, whereas large continental nuclei, such as the Hehuai Continental Nucleus, were formed 2500 my ago (in
the Fuping Movement). The major part of the paraplatform was essentially
consolidated 2000 my ago (in the Wutai Movement), but the paraplatform
was finally shaped 1700 my ago (in the Zhongtiao Movement). The peripheral zones of Bayan Obo and other areas were formed even later.

On the basis of the results of aeromagnetic surveys (Fig. 19), Zhu Ying
(1979) has recently grouped the basement tecto-lithofacies for the major
part of the Sino-Korean Paraplatform into the following three magnetic field
areas.

1. *The Baotou-Yinchuan Area*

This embraces a triangular area lying south of the Linhe-Zhangjiakou
line, east of the Helan Mt. and northwest of the line connecting Zhangjia-
kou, Datong, Jingbian and Guyuan. Its characteristic is that the negative
magnetic fields cover a very large area, which reflects the feature of the magnetic fields over the Archean Sanggan Group in Baotou, Shizuishan and

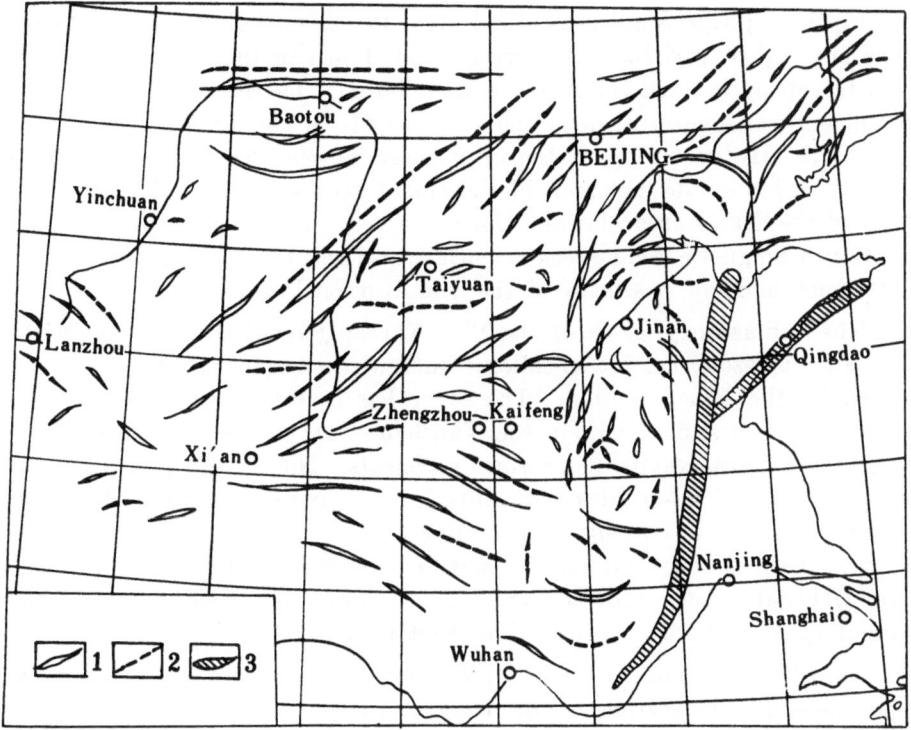

Fig.19. Diagramatic map showing the distribution of the magnetic anomaly trends in the Sino-Korean Paraplatform (After Zhu Ying 1979).

1 axis of positive anomaly; *2* axis of negative anomaly; *3* strong linear anomaly zone distributed within deep fracture belt

other places.

2. *The Taiyuan-Xi'an Area*

This refers to a triangular area lying southeast of the Datong-Guyuan line, west of the Beijing-Handan-Queshan line and north of the line connecting the Xi'an, Tongguan, Luoning, Lushan and Queshan. In the area, the rising positive magnetic field zones and the gentle negative magnetic field zones tend to predominate. The positive anomaly zones are characterized by three sets of very prominent trends, of which the NE- and E-W sets may probably correspond to the principal directions of the magnetic fields of the early Proterozoic Wutai Group and the Archean Fuping Group, while the NNE set may be related to granodiorite terrains of Mesozoic age.

3. *The Tianjin-Bengbu Area*

This lies to the east of the Taiyuan-Xi'an magnetic field area, west of the Tancheng-Lujiang fracture, south of Qinhuangdao (including the southern part of the Bohai Sea) and north of the Hefei and Queshan[5]. This is

---

5 Slight Changes of the boundary here are made by the authors.

an elliptical, positive magnetic field area, in which the magnetic field directions are highly variable. Its core is located just in the area where the Taishan Group in western Shandong is distributed. Therefore, it may represent an ancient continental nucleus made up mainly of the Taishan Group.

As the Hutuo Group of the uppermost Lower Proterozoic and its corresponding low-grade metamorphic basement rock series and the sedimentary cover of the Sino-Korean Paraplatform are very weakly magnetic, the abovementioned magnetic field areas essentially reflect the tectonic configuration of the metamorphic basement of the Sino-Korean Paraplatform.

The sedimentary cover of the Sino-Korean Paraplatform begins with the Upper Proterozoic, i.e. the so-called Sinian of the northern China type represented by the Jixian section. The cover has a total sedimentary thickness up to ten thousand meters in the Yanshan area; and is therefore often regarded as a parageosyncline by some geologists. Now it has been ascertained that the nearly 1000-m-thick sediments in this "parageosyncline" have a time range from 1950 to 850 my, with an average sedimentation rate of only 9 m/10^6 yr. Besides, the dominance of neritic, terrigenous fragments and magnesian carbonate rocks and the presence of abundant stromatolites in the sedimentary formations also fully show that it is a sedimentary cover.

The strata equivalent to the Sinian System of the southern China type represented by the section in the Yangtze Gorges are absent in most areas of the Sino-Korean Paraplatform; they are only found in western Henan and Huainan on the southern margin of the paraplatform, as well as eastern Liaoning, and the stratigraphical correlations there have not yet been ascertained up to now.

In the platform cover, the Cambrian and Ordovician systems are best developed. They are widespread, dominated by neritic carbonate formations, their lithofacies and thickness being persistent, generally 1500 m or so.

The strata of the Upper Ordovician to Lower Carboniferous are commonly absent except for the transition belt on the western margin of Ordos (the Ordos Platform-Margin Fold Belt). This is a very important salient feature of the Sino-Korean Paraplatform.

The Middle and Upper Carboniferous and Lower Permian are represented by littoral and continental coal-bearing formations, which make up the most important coal measures in northern China.

In the Late Permian, large-scale continental sedimentary basins began to form on the Sino-Korean Paraplatform, on which were deposited the Upper Permian to Middle and Lower Triassic red formations (originally called the Shiqianfeng Group).

After the Indosinian Movement (Late Triassic), the Sino-Korean Paraplatform entered the stage of development of a continent-margin activization belt, as is the case with other regions in eastern China, forming the large Mesozoic continental sedimentary basin of Ordos. During the Yanshanian Cycle, large-scale volcanic eruption of mainly acid and intermediate magmas and intrusion of granitic magma occurred in Yanliao, Shandong, Inner

Mongolia and other regions, accompanied by intense folding and fracturing of the platform cover, thus bringing about a radical change in the original tectonic framework of the Sino-Korean Paraplatform. In the Cenozoic, fault-block uplift and subsidence movements dominated, accompanied by eruption of basaltic magma, thus giving rise to such famous down-faulted basins as the Fenwei and Wuyuan-Huhhot Basins and the gigantic North China-Bohai Sea Epicontinental Basin.

Here the Alxa Platform Uplift needs a special explanation. Some Chinese geologists consider that it is not a platform but a part of a Paleozoic geosyncline. Here we still believe that it represents a platform uplift. This is supported by the following lines of evidence: (1) The Longshoushan Group in the Longshou Mt. area is intruded by 1786- and 1719 -my -old granites; on it lies unconformably the Dunzigou Formation of the late Proterozoic low-grade metamorphic rock series and further above lie tillites of the Sinian Hanmushan Formation; (2) there occurs the Corridor Transition Belt between Alxa and the Qilian Geosyncline, which clearly shows the phenomenon of tectonic transit on from a stable platform to a mobile geosyncline; (3) the pre-Sinian metamorphosed basement has been exposed in the Helan Mt. area, which is overlain by the platform-type Jixian System (the Helan Group, containing stromatolites) and further above by tillites of the Muguan Formation; (4) there occurs the high-grade metamorphic Alxa Group north of Jartai, which is intruded by 1365-my-old gabbro and overlain by the Late Proterozoic metamorphic rock series; and (5) The northern margin of Alxa is separated from the Jianshan-Hinggan Geosynclinal Fold Region by a deep fracture, and both sides of the deep fracture have utterly different histories of tectonic development. The time of the formation of the platform-uplift basement may be assigned to the Yangtze Cycle. To the west of Alxa a Proterozoic uplift of the isthmus type connects the Sino-Korean Paraplatform with the Tarim Platform.

III.3   THE YANGTZE PARAPLATFORM

The Yangtze Paraplatform includes almost the entire Changjiang (Yangtze) River Valley, extending from eastern Yunnan to Jiangsu and the southern part of the Huanghai (Yellow) Sea. This is a platform formed in the Yangtze Cycle at the end of the Late Proterozoic. It is separated on the southwest from the Sanjiang (Three-River) Fold System by the Jinsha River Honghe (Red R.) Deep Fracture, on the northwest from the Songpan-Garzê Fold System by the Longmen Mt. Deep Fracture, on the north from the Qinling Fold System by the Chengkou-Fangxian and Xiangfan-Guangji Deep Fractures, and from the Sino-Korean Paraplatform by the Tancheng-Lujiang and Jiashan-Xiangshui Deep Fractures, and on the southeast from the South China Fold System by the Nanpan River Xupu-Sibao and Jiangshan-Shaoxing Deep Fractures.

The Proterozoic basement rock series making up the Yangtze Paraplatform is mostly exposed on the margins of the paraplatform, for example, the

Kunyang Group, Huili Group and Kangding Complex at the Kam-Yunnan Axis on the western margin; the Baoxing Complex and the Pengguan Complex in the Longmen Mt. and the Huodiya Group and the Hannan Complex on the Yingzuiyan Uplift in Hannan on the northwestern margin; and the Lengjiaxi Sibao and Banxi Groups on the southeastern margin. Within the platform, the basement is only observed at a few localities, its exposed area being small, which is exemplified by the Kongling Group on the Huangling Anticline, the Shennongjia Group at Shennongjia, the Emeishan Granite in Mount Emei and the Banxi Group and the Fanjingshan Group in Mount Niushou, eastern Yunnan, and in Mount Fanjing, eastern Guizhou.

On the basis of the geological and geophysical data available, three basic types of basement have been distinguished for the Yangtze Paraplatform, namely: (1) the central Sichuan type, (2) the Jiangnan type, and (3) the Kunyang type. The basement of the central Sichuan type is distributed in central Sichuan, central and western Hubei and the southern Huanghai Sea, and its characteristics lie in the fact that the basement rocks are highly metamorphosed and rigidified. The Jiangnan type occurs in the Jiangnan Uplift (Axis) and its surrounding areas; there is a gradational or not abrupt transformation process between the basement and the cover, and the degree of rigidity of the basement is very low. The Kunyang type is mainly represented by the Kam-Yunnan Axis Region in eastern Yunnan. Its basement is composed of low-grade metamorphic rocks and its degree of rigidity is intermediate between the central Sichuan and Jiangnan types.

The study of the basement of the central Sichuan type is not yet adequate at present, as it is mostly buried beneath the Paleozoic, Mesozoic and Cenozoic deposits. From geological and geophysical data available, some people infer that it might be the oldest component part of the basement of the Yangtze Paraplatform, representing an ancient continental nucleus formed at the initial stage of the paraplatform, but further study is needed to confirm this inference. One point that can be affirmed is that all these areas underwent strong tectono-magmatic activity during the Jinning and Chengjiang Movements, which is supported by the isotopic age determinations of the Emeishan Granite and the Huangling Complex and by the deep core drilling data from central Sichuan.

The formation of the Kunyang-type basement underwent two stages. The early stage is an eugeosynclinal stage during which the Dahongshan Group and its corresponding strata were deposited; the late stage is a miogeosynclinal stage during which the deposition of the Kunyang Group took place. After the Jinning Movement, the geosyncline was transformed into a platform.

The formation of the Jiangnan-type basement also involved two stages. The early stage is an eugeosynclinal stage during which the Lengjiaxi and Fanjingshan groups were deposited; they were folded for the first time during the Wuling Movement (Dongan Movement). The late stage is a miogeosynclinal stage during which the Banxi Group was deposited. Then the geosyncline was transformed into a platform after the Xuefeng Movement. The characteristics of the Xuefeng Movement show that it is not an intense folding movement. Therefore, structurally the Banxi Group in western Hunan, eastern Guizhou and northern Guangxi is generally quite gentle, and there is no remarkable angular unconformity between it and its cover. It is generally considered that the Lengjiaxi Group is equivalent to the Sibao Group, but recently Qiao Xiufu et al. (1980, unpublished data) suggested that the Sibao Group in northern Guangxi is another sequence of eugeosynclinal deposits younger than the Lengjiaxi Group (and Fanjingshan Group), and considered that it represents the second sequence of eugeosynclinal deposits south of the ancient continental nucleus of central Sichuan that accreted southeastwards during the Proterozoic.

The Xuefeng Movement tends to be correlated with the Jinning Movement; but Liu Hongyun et al. (1973) presented a quite different view. He maintained that the Xuefeng Movement is equivalent rather to the Chengjiang Movement than to the Jinning Movement. If this is true, then the three types of basement of different nature within the Yangtze Paraplatform may represent the products of three different stages in the formation of the basement of the Yangtze Paraplatform. Of the three types, the central Sichuan type is the oldest, having the highest degree of rigidity; the Kunyang type comes next, with an intermediate degree of rigidity; the Jiangnan type is the youngest, its degree of rigidity being the lowest. Further work is required to prove all these ideas.

The sedimentary cover of the Yangtze Paraplatform is very well developed. The first cover, of the Sinian to Silurian in age, is widespread throughout the platform. Its lower part consists of molasses and tillites represented by the Chengjiang Sandstone and the Nantuo Tillite, accompanied by rather strong volcanic activity in some areas. The middle part, from the Upper Sinian to Ordovician, is mainly marked by a neritic carbonate formation, in which the Sinian System and the Lower Cambrian are the most important known phosphatic sequences in China. The upper part, from the Upper Ordovician to Silurian, is represented by a sequence of graptolitic shale and argillaceous shale. In the Wuling Depression, western Zhejiang and southern Anhui near the South China Caledonian Geosyncline, the cover is very thick, and shows a gradation into the South China Caledonian Geosyncline. The Upper Ordovician in western Zhejiang and southern Anhui exhibits distinct features characteristic of turbidites. Actually, it had become a portion of the South China Caledonian geosyncline by that time. The second cover is composed of the strata from the Devonian to Middle Triassic. The lower part of the cover, from the Devonian to Early Carboniferous, is mainly distributed on the western margin (in the Longmen Mt., Yanyuan and Lijiang) and the southern margin (eastern Yunnan, southern and eastern Guizhou, western Zhejiang and southern Anhui) of the platform and the Lower Changjiang valley, and is absent or very thin in other wide areas. It is best developed in the Longmen Mt., Lijiang and southern Guizhou, and its maximum thickness may reach 4000—6000 m in the Longmen Mt. The Lower Devonian is composed of a clastic series represented by the Pingyipu Formation, while the Middle and Upper Devonian and Lower Carboniferous consist of carbonate formations. The middle part of the cover, from the Middle and Late Carboniferous to the Early Permian, is marked by carbonate formations, not very thick, with persistent lithofacies, almost wideseperead all over the region (the Middle and Upper Carboniferous are lacking in central Sichuan and other areas). In the Late Permian, continental-littoral coal measures were extensively developed, and there are large-scale basalt outpourings in western Sichuan, eastern Yunnan and western Guizhou. The upper part of the cover is composed of Triassic sequences. The lower part of the Lower Triassic consists of littoral-neritic red beds represented by the Feixianguan Formation in the western half of the paraplatform, while in the eastern half it consists of a carbonate formation represented by the Daye Limestone. The upper part of the Lower Triassic and the Middle Triassic are composed of a gypsum- and salt-rich lagoonal carbonate formation and a red formation, evidently representing the regressive sediments deposited when the sea-water gradually retreated from the Yangtze Paraplatform.

Since the Late Triassic, the Yangtze Paraplatform, like most areas of eastern China, has entered the stage of development of a continent-margin activization belt. The deposits occurring during this stage are all of continental origin except for the western part of the platform, where the Upper Triassic is characterized by marine and paralic deposits, as well as for the middle and lower reaches of the Changjiang River valley where from the Late Cretaceous to Cenozoic there occurred several short sea transgressions and floodings. With regard to Mesozoic deposits, in the western half of the platform, the lower part of them is represented by the Late Triassic to Early Jurassic coal-bearing formations (partly coal-bearing molasses), including the Xiangyun coal measures and the Baitianba Formation; the upper part of them, from Jurassic to Cretaceous, is composed of the red beds and molasse formations, forming the famous Sichuan Red Basin and the central Yunnan Red Basin. In the eastern half of the platform, the characteristics of the Mesozoic deposits in the lower Changjiang River valley consist in the development of the continental volcanic series in Late Jurassic to Early Cretaceous times. Cenozoic deposits are mainly distributed in the eastern part of the platform, consisting of the gypsum- and salt-bearing red formations and oil-bearing formations.

Since the Late Triassic, the original tectonic features of the Yangtze Paraplatform have experienced very strong modifications. The first modification was caused by the Indosinian Movement. It was strongly felt on the western margin of the platform and the lower Yangtze area, generating strong folding and overthusting in the Longmen Mts. area and being characterized by intermediate and acid magmatism in the Kam-Yunnan Axis. The second modification was produced by the Yanshanian Movement which is the strongest tectonic movement that the Yangtze Paraplatform has ever been subjected to since the Paleozoic. It affected the entire platform and gave rise to an extensive platform-cover fold belt. In the lower Yangtze area, this movement was accompanied by strong magmatism, mainly acid and intermediate in composition. The third modification was generated by the Himalayan Movement which brought about an overall folding and uplifting of the western half of the entire platform, including the Sichuan Basin and the central Yunnan Basin, and formed the final shape of the Longmen Mt. Nappe Belt. At the same time, the eastern half of the platform sank intensely, leading to the continuous development of the Jianghan and northern Jiangsu-southern Huanghai Sea basins that were formed in the late Yanshanian stage.

The Yangtze Paraplatform may be further divided into the following second-order tectonic units:

1. *The Platform-Margin Fold Belts on the Northern Margin of the Yangtze Paraplatform*, which refer to the Yanyuan--Lijiang and Longmen-Daba Platform-Margin Fold Belts. The former represents a platform-margin depression belt of Paleozoic to Triassic ages, and also a transitional belt between the Yangtze Paraplatform and the Songpan-Garzê Geosyncline. Since the Mesozoic, it has been strongly affected by the Indosinian, Yanshanian and, particulary, Himalayan Movements, thus forming a rather strongly deformed platform-margin belt and fracture belt. The latter was originally mainly a Paleozoic, particularly early Paleozoic, platform-margin depression belt. It was totally folded and became a platform-margin fold belt during the Yanshanian Movement.

2. *The Kam-Yunnan Axis,* which used to be an uplift belt during the period from the Sinian to Middle Triassic, and also an important tectono-magmatic metallogenic belt on the Yangtze Paraplatform during the Variscan and Indosinian cycles. After the Indosinian movement, it was transformed from an uplift belt into a down-faulted basin——the central Yunnan Basin. It was then folded and uplifted again during the Himalayan Movement.

3. *The Sichuan Platform Depression,* which used to be a relatively uplifted zone on the Yangtze Paraplatform during the Paleozoic, with the Devonian and Carboniferous systems being absent. It was transformed into a major depression after the Indosinian Movement, and folded and uplifted to the full in the late Yanshan'an Movement, particularly after the Himalayan Movement. The superficial folding was very smooth in central Sichuan and other areas as the basement there was rigid.

4. *The Upper Yangtze Platform Fold Belt,* which includes the original Bamianshan Fold Belt and the Yunnan-Guizhou Fold-Fracture Region. During Paleozoic—Middle Triassic times, the marine sedimentary cover was well developed. It was folded throughout during the Yanshanian Movement, thus forming a platform-cover foldbelt.

5. *The Lower Yangtze Platform Fold Belt,* which represents a Paleozoic-Triassic depression belt that has undergone the Indosinian and Yanshanian polycyclic tectono-magmatism since the Mesozoic, thus forming a structurally complex platform-cover fold belt.

6. *The Jiangnan Platform Uplift,* which was formerly termed the Jiangnan Axis. This is a long-term active uplift belt on the southeastern mar-

gin of the Yangtze Paraplatform, and tectonically displays a distinct nature of transition. Therefore, some people regard it as a marginal uplift belt of the Yangtze Paraplatform, while others consider it as a marginal geoanticlinal belt of the South China Caledonian Geosyncline. It has been subjected to several modifications by the Indosinian, Yanshanian and Himalayan Movements at a later stage.

7. *The Jianghan Fault Depression,* which is a continental down-faulted basin that began to develop in the Cretaceous after the middle Yanshanian phase.

8. *The Northern Jiangsu Fault Depression* (including the Northern Huanghai Sea), which represents a continental down-faulted basin that began to develop in the Cretaceous after the middle Yanshanian phase.

9. *The Western Zhejiang-Southern Anhui Platform Fold Belt,* which represents a depression belt that was formed from Paleozoic to Triassic times. Its characteristics lie in the fact that the lower Paleozoic, particularly the Ordovician and Silurian, exhibits salient features of geosynclinal sediments. So some people tend to assign it to the South China Fold System.

III.4 THE TARIM PLATFORM

The Tarim Platform, rhombic in form, lies to the south of the Tianshan Mts. and to the north of the Kunlun Mts. The broad area of the platform is covered by Cenozoic deposits except its peripheries, where the basement made up of pre-Sinian metamorphic rocks and the Paleozoic sedimentary cover are found to be exposed.

According to the Regional Stratigraphic Table of Xinjiang, the oldest strata in the Tarim Platform are the Daglagh Brak Group.. This group, 1000 m thick, consists mainly of various kinds of gneisses, which are unconformably overlain by the Early Proterozoic Shingdi Tagh Group. According to its contact with its overlying strata, the age of the group is assumed to be Archean. The Shingdi Tagh Group, 3500 m in thickness, is composed of quartz-schist, mica-schist and schistose metamorphic sandstone, which is unconformably covered by the Yanggi Brak Group. The Yanggi Brak Group, 2500 m thick, is represented by intermediate- and low-grade metamorphic neritic clastic rocks, which is overlain by the Ailchikan Group. The Ailchikan Group, 2500 m thick, consists of marble with chert nodules or banded marble and dolomitic marble. As it contains such stromatolites as *Kussiella kussiensis* (Maslov) Krylov, the group is assigned to the Jixian System (late Precambrian). Overlying conformably the Ailchikan Group of the Jixian System (late Precambrian) is the Pargang Tagh Group of the Qingbaikou System (late Precambrian). The lower part of the latter is composed of low-grade metamorphic clastic rocks; while its upper part, of crystalline limestone and dolomite, 1500 m thick, containing stromatolites, such as *Katavia paergangensis* Kao, *Baicalia* sp. and *Inzeria sinkiangensis* (Mou)

Kao. The above-mentioned systems are all metamorphosed and tightly folded, forming the basement of the Tarim Platform. The basement is unconformably overlain by the Sinian-Paleozoic marine deposits and Mesozoic-Cenozoic continental deposits. From several unconformities, it is known that the Tarim Platform was formed as a result of polycyclic tectonic movements and that its basement was finally consolidated at the end of the Yangtze Cycle, i.e. it is a platform of the Yangtze Cycle. The movement that took place at the end of the Yangtze Cycle as referred to here is called the Tarim Movement by Gao Zhenjia (1978) which shows that this movement is of universal significance to the formation of the Tarim Platform.

The Sinian System, which is best developed in the Kuruk Tagh area, has also been studied in detail. The Sinian Kuruk Tagh Group rests unconformably on the Pargang Tagh Group of the Qingbaikou System. According to Gao Zhenjia et al. (1979), the Kuruk Tagh Group is of littoral-neritic origin, composed of clastic rocks, volcanic rocks and three horizons of tillites containing an abundance of algae, with a total thickness of 3000—7800 m. The Cambrian and Ordovician systems are composed of limestones, the Lower Silurian, of sandstone and siltstone; the Devonian System, mainly of sandstone and conglomerate; the Carboniferous-Permian sequence, mainly of carbonate rocks and clastic rocks. The Mesozoic and Cenozoic erathems consist of continental deposits, except on the western margin where the Upper Cretaceous and Lower Tertiary are characterized by marine deposits.

The platform-cover deposits were folded and fractured as a result of the Variscan and Alpine movements. Whether or not the Caledonian Movement has affected the platform is still open to question. But from the fact that the Upper Ordovician is absent in a number of areas, it follows that the Early Caledonian Movement has affected the platform. The Early Silurian is identified according to the graptolite assemblage, while the corals therein take on the character of the Middle and Late Silurian; so whether the Middle and Upper Silurian are lacking remains to be studied. From the fact that the Devonian is mostly of continental origin, it may be, however, concluded that the Late Caledonian Movement also affected the platform. The Middle Carboniferous lies unconformably on the Devonian System (the Akqi stratigraphic subprovince); the Upper Carboniferous, on the Middle Carboniferous (the Akqi stratigraphic subprovince); and the Upper Permian, on the Lower Permian (the Baicheng stratigraphic subprovince). These contact relationships may indicate that the Variscan Movement formerly brought about foldings and fracturings of the Paleozoic cover, and eruption of basalts in some areas. The Variscan Movement had greatly changed the aspect of the platform. A vast expanse of sea-water became a stretch of land, with the Triassic System being absent in vast areas except in some places on the northern margin of the platform. The Jurassic System is also distributed only on the peripheries. The Cretaceous System is distributed over broad tracts in the southwest. The Lower Tertiary spreads all over the region, and the Upper Tertiary is even thicker. From the above facts it follows that the Tarim Platform began to transform into a depression region in the Tertiary.

The basement of the depression is uneven. It is separated into the northern and southern depressions by a median uplift. The amplitude of subsidence of the two depressions increases towards the piedmonts. Geophysical data and Landsat photographs show that the interior of the depression, like its margins, is also cut by the WNW- and ENE-trending fractures. It must be pointed out that a nearly E-W-trending concealed fracture cuts through the platform, and joins the Shulehe Fracture to the east and penetrates the Kunlun Mts. to the west. This fracture was first determined in the light of the Landsat photographs, and was later found to exist also from geophysical data.

Owing to the northward compression of the Indian plate, the Himalayan Movement in the Tarim Platform and its surrounding areas took the form not only of uplift and subsidence of fault blocks, but also made the older strata thrust over the Upper Tertiary, with the latter forming overturned folds.

The Tarim Platform may be divided into eight second-order units. They are: the Kuruk Tagh Fault Uplift, the Kalpin Fault Uplift, the Tikanlik Fault Uplift, the Northern Platform Depressions, the Central Platform Depression the Southern Platform Depression, the Kuqa Piedmont Depression, and the Kunlun Piedmont Depression.

The Kuruk Tagh Fault Uplift. It is well known for the exposure of a relatively complete succession of strata. The strata, especially those of the Sinian and the Cambrian-Early Silurian times, are characterized not only by a greater thickness, but also by multilayered volcanic rocks. Some of its features resemble those of geosynclinal type. This is a problem that deserves further discussion.

The Kalpin Fault Uplift. The basement rocks are not completely exposed, and the Paleozoic cover is rather thin, but the sediments tend to thicken towards the South Tianshan Mts.

The Tikanlik Fault Uplift. The Precambrian strata are widespread and all metamorphosed. Nephrite is known to occur in them. The Paleozoic strata are only exposed in small amounts. The sedimentary facies resembles that of the Kalpin Fault Uplift.

Based on the sedimentary and structural characteristics, the Tarim Platform is considered to be a favourable area for prospecting for oil. The authors hold that it will be proved to be a better one than the Songliao Depression in this respect.

III.5 THE SOUTH CHINA SEA PLATFORM

The South China Sea Platform is a historical tectonic unit and most parts of it are no longer in existence today. The main grounds for suggesting this paleotectonic unit are as follows:

1. Near Yaxian County, in the extreme south of Hainan Island, the Middle Cambrian to Ordovician strata are represented by a phosphorus- and

manganese-bearing siliceous carbonate formation, graptolitic shale and a quartz-sandstone formation, containing an abundance of trilobite and graptolite fossils. Evidently they are different from the flysch and flysch-like deposits in the Caledonian South China Geosyncline;

2. Drilling has revealed that the Neogene deposits are underlain by granite gneiss, quartz-mica schist and gneissic granite on Yongxing Island of the Xisha Islands. Their isotopic age (Rb-Sr whole-rock isochron age) is 627 my, which represents the age of primary deposition, corresponding to the latest Precambrian. Their metamorphic age is 77 my (unpublished data from the Ministry of Geology and Mineral Resources, and the Ministry of Petroleum Industry.); and

3. On the Indosinian Massif west of the Xisha Islands, the oldest cover on the Precambrian crystalline basement is represented by Middle Cambrian flat-lying quartzite (The *Geological Map of Asia*, scale 1:5000,000, compiled by the Chinese Academy of Geological Sciences 1975).

III.6 THE SAYAN-ERGUN GEOCYNCLINAL FOLD REGION

The main part of this fold region is located within the territories of the Soviet Union and Mongolia, and only a small part of it lies within the territory of China, i.e. the Ergun Fold System.

The Ergun Geosynclinal Fold System is located in northern Heilongjiang. On the south it is separated from the Greater-Hinggan Eugeosynclinal Fold Belt by the Derbur Deep Fracture; southwestwards it crosses the border line and is connected with the Mongolian-Transbaikalian Fold System and the Central Mongolian Fold System in Mongolia, and northeastwards it extends into the Soviet Union and is connected with the Bureya Massif. It is distributed largely in a northeast direction.

Ergun is a geosynclinal fold system of the Xingkai Cycle, but the strata are distributed discretely as they are penetrated by Variscan granites and covered extensively by Yanshanian velcanic rocks. In ascending order, the strata of the Xingkai Cycle consist of (1) the Sinian-Lower Cambrian Jiageda Group, (2) the Lower Cambrian Ergun Formation, and (3) the Lower Cambrian Anniangniangqiao Formation. The three sequences, nearly 5000 m thick, are all metamorphosed and display conformable contacts with each other. They are overlain unconformably by the Lower Devonian. The Jiageda Group, 2500 m thick, may be divided into the upper and lower parts: the lower part is marked by mica-quartz-schist, and the upper part by greenschist; it contains spores of *Leiominusoula*. The Ergun Formation, 1100 m thick, consists mainly of marble, dolomitic marble and dolomite, containing phosphorus mineralization and spores of *Leiominusoula*. The Anniangniangqiao Formation, more than 1200 m thick, mainly includes greenschist, with acid lavas and tuffs (Compilation Group of the Regional Stratigraphic Table of Heilongjiang Province 1979).

The above-mentioned Sinian-Lower Cambrian geosynclinal deposits were folded as a result of the Xingkai Movement at the end of the Early Cambrian, and were inlaid on the margin of the Siberian Platform. The Paleozoic geosyncline was detached from the Ergun Fold System and moved southwards, while Ergun had been in a state of uplift for a long time. Transgression occurred in the Late Paleozoic, but the deposits show the nature of a cover.

The exposure of large amounts of granites of Variscan time shows that the Variscan Movement had exerted strong effects on Ergun, so that many strata were swallowed up by granites and the tectonic pattern generated by the Xingkai Movement was destroyed. The Yanshanian Movement caused Ergun to undergo an additional strong modification, and thus the NEN-trending fractures and volcanic rock belts became striking tectonic landscapes.

III.7 THE TIANSHAN-HINGGAN GEOSYNCLINAL FOLD REGION

Lying north of the Tarim Platform and the Sino-Korean Paraplatform, the Tianshan-Hinggan Geosynclinal Fold Region makes up a part of the gigantic Central Asian-Mongolian Geosyncline, comprising the Altay, Junggar, Tianshan, Inner Mongolian-Greater Hinggan and Jilin-Heilongjiang fold systems.

III.7.1 The Altay Fold System

The Altay Geosynclinal Fold System is located in the extreme north of Xinjiang, distributed in a northwest direction, parallel to the Altay Mts. It extends northwestwards across the borderline and is connected with the mountainous Altay Fold Belt and the mining Altay Fold Belt in the Soviet Union, and extends southeastwards across the borderline and is connected with the Mongolian Altay Fold Belt. On the south it is separated from the Junggar Fold System by the Ertix Deep Fracture.

The Altay Geosyncline within the territory of China is composed of two utterly different types of sedimentary facies. The Lower Paleozoic belongs to the Bayan Har-type miogeosyncline, while the Upper Paleozoic to the Tianshan-type eugeosyncline (for the meanings of the Bayan Har-type miogeosyncline and the Tianshan-type eugeosyncline, see Chap. IV). The oldest strata exposed are the Middle and Upper Ordovician, represented by metamorphic rocks of different grades. The protoliths consist of sandstone, siltstoae, phyllite with marble, lenticular conglomerate and siliceous shale, over 6000 m thick. A comparison with the data available from the Soviet Union (Volkov 1966) shows that the strata belong to the flysch formation. The Silurian System is still represented by metamorphic clastic rocks, also nearly 6000 m thick. This shows that the entire Lower Paleozoic belongs to miogeosynclinal deposits. The Devonian System consists of acid volcanic rocks, volcaniclastic rocks and clastic rocks with limestone, and there are appreciable changes in lithofacies, but it essentially belongs to eugeosynclinal deposits, more than 6000 m thick. Geosynclinal deposition concluded at the end of the Devonian. Afterwards, the geosyncline might have been in a state of uplift for a short time, and there were deposited the 500—700 m-thick uppermost Early Carboniferous strata, composed mainly of acid volcanic rocks, limestone and clastic rock, which belong to cover deposits.

The Chinese Altay is located in an area linking the Soviet Union Altay with the Mongolian Altay, which is the link part of the southeastward migration of the structure. In the mountainous Altay in the Soviet Union, the Cambrian and Ordovician Systems are exposed over vast areas, and the early Caledonian Movement (terminal Early Ordovician) brought about folding and uplift of its northwestern part (Volkov 1966); the Cambrian and Ordovician strata have not been found yet in the Chinese Altay. The Silurian System is scattered within the territory of the Soviet Union near China; and is also not much exposed at the northwestern end of the Chinese Altay; but it is relatively widespread towards the southeastern end and much more widespread within the territory of the Mongolian People's Republic (Marinov 1973). From the above it may be seen that the particular terrain is gradually folded or uplifted from northwest to southeast along the Altay, exhibiting a longitudinal structural migration.

Granites in the Altay are mostly of Variscan age, but there are different views on their origin. One proposal is that they are the products of magmatic activities, while another considers that they are the products of granitization, as a transitional relationship may be observed between granite and its surrounding strata in some places. According to the data available, we are inclined to believe that both kinds of granite may occur in the Altay.

Some people suspect that the high-grade metamorphic rocks in the Altay, such as gneiss, might be of Precambrian origin, but there is no evidence for this at present. According to the data from neighbouring areas and an analysis of tectonic settings, we should think that there is little possibility of that. From the Sinian to the Early Cambrian, most parts of the Altay had been in an oceanic environment, which is evidenced by a considerable thickness of strata containing a large amount of spillite. The Xingkai Movement caused the geosyncline to be differentiated: the Salair area on the southern margin of the Siberian Platform rose, the Altay area transformed from an eugeosyncline into a miogeosyncline, and the Junggar area continued to be in an oceanic environment.

III.7.2 The Junggar Fold System

The Junggar Fold System is located in a vast area south of the Altay Mts. and north of the Tianshan Mts. Westwards, it crosses the border line and is connected with the Junggar-Balkhash Geosyncline; and eastwards, it crosses the border line and is connected with the southern Mongolian Geosyncline.

The Junggar Geosynclinal System used to be a major part of the Central Asian-Mongolian Ocean. It inherited the Late Proterozoic oceanic environment and continued to receive deposits in the Early Paleozoic, thus forming a Qilian-type eugeosyncline. Through the Caledonian Movement, the geosyncline was differentiated to form a Late Paleozoic eugeosynclinal belt of dominantly andesitic volcanic rocks and volcaniclastic rocks and a Paleozoic miogeosynclinal belt of terrigenous clastic rocks and carbonate rocks. The geosynclinal deposition concluded in the terminal Middle Variscan time (terminal C_1). The Middle and Upper Carboniferous and Permian Systems are composed mainly of continental volcanic and clastic rocks, while the Middle Carboniferous marine strata occur only as interbeds in some places. The Variscan Cycle gave rise to major depressions and minor intermontane depressions; the Yanshanian and Himalayan Movements caused the fold belts to be extremely fragmented and outcrop as fault blocks.

At present the only known oldest strata exposed in the Junggar Geosynclinal System are the Middle Ordovician ones. In the light of the fact that Sinian-Cambrian eugeosynclinal formations are exposed in the Kingiss-Tarbagatai of the Soviet Union (Esenov et al. 1972), it is inferred that at that time the Junggar geosyncline might be in an oceanic environment. During the Middle and Late Ordovician, an oceanic environment prevailed in the Mayile Mountain north of Ebinur Lake, forming an ophiolite suite, while in the Sarburt Mountain southwest of Ulungur Lake, an island chain occurred, followed by deposition of calcareous sandstone, leopard limestone, tuff, andesite porphyrite and reef libestone with a total thickness of more than 1000 m. The post-Silurian or pre-Middle Devonian Late Caledonian Movement led to folding of the eugeosyncline in the Mayile Mt. as well as of the Early Devonian and pre-Early Devonian strata in the Sarburt Mt. (considered by some to be disconformities, as the contacts are covered.) which are overlain by continental molasse formations of Middle Devonian age. This movement resulted in intense differentiation of the geosyncline: andesitic volcanic rocks and volcaniclastic rocks formed in the Mayile Mt., while clastic rocks with occasional volcanic materials of the geosyncline-platform transition type were deposited in the Sarburt Mt., and the geosyncline was pushed northwards to the Sawur Mt. In eastern Junggar, the unconformities caused by this movement have not been observed directly as they were destroyed by subsequent fractures, but distinct indications of differentiation of the geosyncline can be seen: euformation belts composed of volcanic rocks and volcaniclastic rocks and mioformation belts composed of clastic rocks and limestone are distributed alternately in plan. Owing to the differentiation of the geosyncline the sediments with volcanic materials and the normal sediments without volcanic materials occur in alternating narrow zones. Jiang Chunfa called ed the former the euformation belt and the latter the mioformation belt. This shows that the late Caledonian Movement also had great effects on eastern Junggar. From the Early Carboniferous on, tectonic movements occurred frequently. Four unconformities have been observed from Lower Carboniferous to Lower Permian strata, which indicates that at least four tectonic movements took place during this period.

Rock masses of different sizes and different kinds originated in response to various tectonic movements. In addition to the rocks exposed in the Mayile Mt., which have been determined as ophiolites, the ultrabasic rocks at

Hongulerun in the Sarburt Mt. and the ultrabasic rocks in the Almantai and Kalameiri belts in eastern Junggar might also be ophiolites (When this paper was at the finalization stage, Zhang Chi from the Xinjiang Institute of Geological Sciences, who has been engaged in the study of ultrabasic rocks for more than 20 years, told us personally that ultrabasic rocks in a few places there really belong to ophiolites). The Almantai ultrabasic belt might be linked with the ophiolitic melange belt (Rotarash et al., 1974) in the Char Fold Belt (Esenov, 1974) of the Zaisan Geosyncline in the Soviet Union. The Kalameili ultrabasic belt extends southeastwards to its border with the Northern Tianshan Mountains, along which ultrabasic rocks and Early Carboniferous spilite and radiolarian siliceous rocks are exposed (unpublished data of the Regional Geological Survey Party of Xinjiang 1977). The two ultrabasic belts in eastern Junggar are most likely to be the final closing lines of the Junggar oceanic basin. Granite belts are usually exposed in the vicinity of ultrabasic belts. Different types of useful minerals related to different magmatic rocks are found to occur, among which chromite associated with ultrabasic rocks is of particular importance.

The Junggar Geosynclinal System is characterized by striking polycyclic tectonic movements, distinct structural migration and strong geosynclinal differentiation. The Junggar Geosyncline belongs to the Central Asian-Mongolian structural migration field: the characteristics of its structural migration coincide with those of this migration field. Laterally, the first-order migration is expressed by a southward motion; and second-order migration, by a motion of two uplift belts (the Sarburt and Mayile Mts.) of preceding cycles towards their respective northern and southern sides. Longitudinally, the first-order migration is marked by an eastward motion; and the second-order migration by a westward motion.

Through the Variscan Movement the geosynclinal deposition in the Junggar Geosyncline ended. Some segments of it were abruptly transformed into fold belts, while others were gradually transformed into platforms. In Permian time, large-scale depressions, such as the Junggar Depression, began to come into being, and medium-scale depressions, such as the Santang Depression, originated on the China-Mongolia border. In the Meso-Cenozoic, large- and medium-scale depressions continued to develop, and, in addition, some small-scale depressions, such as the Tacheng Depression, were formed. In the Junggar Depression, oil-bearing horizons being mined are known to occur in the Mesozoic strata, but we believe that the oil-bearing strata of Paleozoic age are likely to be more important.

In Junggar, apart from the NW-trending structure generated by the Caledonian and Variscan movements there are also the NE-, NNW-and ENE-trending structures. The NE structures originated in the Caledonian Cycle and the Variscan Cycle. The NNW-trending structures were formed in the Alpine Cycle. The ENE-trending structures might originate in both the Variscan Cycle and the Alpine Cycle.

The Junggar Geosynclinal Fold System may be further divided into three tectonic units of second order, the western Junggar Eugeosynclinal

Fold Belt, the eastern Junggar Eugeosynclinal Fold Belt and the Junggar Depression.

a) *Western Junggar Eugeosynclinal Fold Belt*

This is characterized by the development of Early Paleozoic ophiolite suites. In the Late Paleozoic andesitic volcanic rocks and volcaniclastic rocks predominated. Owing to the influence of the Darbut Deep Fracture, the fold belt occurs as a series of arcs, convex southwards.

b) *Eastern Junggar Eugeosynclinal Fold Belt*

This is characterized by alternating arrangement of euformation zones and mioformation zones. The ultrabasic zones are the products of late Paleozoic age. Between the Carboniferous and Permian there are by far more unconformities in this belt than in the western Junggar Fold Belt. This belt strikes NW, but the NNW- and ENE-trending fractures may be seen.

c) *Junggar Depression*

This is triangular in form. According to the geological and geophysical data from the Petreleum Administration Bureau of Xinjiang, the depression was formed on the Variscan basement, with its peripheries dissected by fractures. The basement occurs in the form of a slope dipping gently to the south, and becomes abruptly steep near the Urumqi Piedmont Depression and wedges itself into the pediment. The cover is well developed, consisting of terrestrial clastic rocks of Permian to Quaternary ages. It thickens gradually from north to south and suddenly becomes much thicker towards the piedmont depression. Three important unconformities are observed in the peripheral covers of the depression, which indicates the existence of at least three tectonic movements. Around the depression, the Early Jurassic Badaowan Formation is seen to rest unconformably on the Late Triassic Haojiagou Formation. On the western margin the Early Cretaceous Tugufan Group is found to lie unconformably on the Late Jurassic Karase Formation, while on the southern margin the Middle Pleistocene Usu Group is seen to overlie unconformably the Early Pleistocene Xiyu Formation.

The NE-trending structures represented by the Darbut Deep Fracture in the western Junggar Fold Belt have aroused the interset of many geologists. We consider it to be a sinistral shear zone. A series of arcs convexing southwards in Western Junggar mentioned above is characterized by very long western wings, extending across the border line; and very short eastern wings terminating at the fracture. The curvature of the southern arcs is great, while that of the northern arcs is small. From this it may be inferred that the eastern wall moves to the northeast, thus resulting in a series of arcs of this form. Whether this is attributed to the inherited activity of transform faults also deserves further disccussion.

The problem of whether there is a triangular ancient rigid mass in the basement of the Junggar Depression has been a cause of dispute. According to the geophysical data (which have been converted) of the Geological Survey Division of the Xinjiang Petroleum Administration Bureau, as well as to the seismic profiles across the Junggar Depression measured by a department from the Ministry of Petroleum Industry, most workers deny the existence of a Precambrian rigid mass. In the main, we agree with this point of view.

III.7.3 The Tianshan Fold System

The Tianshan Geosynclinal Fold System is located in the northwestern part of China. Its areal extent largely corresponds to that of the Tianshan Mts. Westwards, it extends across the border line and is connected with the Tianshan Mts. of the Soviet Union; eastwards, it is connected with the Inner Mongolian-Greater Hinggan Fold System. In the south the deep fracture on the southern margin of the Tianshan Mts. separates it from the Tarim Platform, while in the north the deep fracture on the northern margin of the Tianshan Mts. separates it from the Junggar Geosynclinal Fold System.

The Tianshan Geosynclinal Fold System within China was formed on the foundation of the Chinese Protoplatform generated in the Yangtze Cycle. The Chinese Protoplatform was disintegrated and transformed into geosynclines after the Xingkai Movement. The fault blocks that have remained after the disintegration of the Protoplatform are exposed intermittently to form the Tianshan Median Uplift Belt, which cut the geosyncline into the Southern Tianshan Miogeosynclinal Fold Belt, the Northern Tianshan Eugeosynclinal Fold Belt, and the Beishan Eugeosynclinal Fold Belt. The resulting Southern and Northern Tianshan Geosynclines and Beishan Geosyncline had further undergene the Caledonian and Variscan Movements and had thus gradually transformed into imbricate fan-shaped fold belts overturned towards both sides. The Yanshanian and Himalayan Movements caused uplift and subsidence of these fold belts, creating high mountains and deep depressions. Therefore, the region is marked by a typical polycyclic geosynclinal fold system.

In both the Tekes to the south of and the Sayram Lake to the north of the Ili Depression, there outcrop the Sinian dolomitic limestone and tillite, which represent the cover of the Chinese Protoplatform. The latest unpublished data obtained by the Regional Geological Survey Party of the Xinjiang Bureau of Geology and Mineral Resources and the Xinjiang Institute of Geological Sciences. Overlying them are the Cambrian System and Lower and Middle Ordovician composed of siliceous phosphatic shale and limestone and sandstone, which represent the formation of transition from a platform to a geosyncline (see the ensuing context) and exhibit the features of the formations of the platform-geosyncline transition type. In the Late Ordovician, typical geosynclinal sediments began to occur owing to the influence of the early Caledonian Movement in the Soviet Union (at the end of the Middle Ordovician). For example, in the northern Tianshan Mts. the Lower and Middle Ordovician consist of clastic rocks and carbonate rocks, more than 1000 m thick, while the Upper Ordovician is represented by thick-bedded limestone, with a total thickness of more than 1800 m. Similar cases

are also found in the southern Tianshan Mts., showing that the differentiation of the geosyncline had become prominent. As the fractures that resulted
from continuous extension of the crust gradually extended deep into the
mantle, basic lavas and jasper rocks occurred in the southern Tianshan Mts.
from Middle Silurian to Middle Devonian times, and a Qilian-type eugeosyncline originated. Afterwards, the eugeosynclinal sediments were replaced
by the miogeosynclinal sediments, and up to the end of the Early Permian
clastic rocks occurred in alternation with carbonate rocks. In the northern
Tianshan Mts., no unconformities have been found between various series of
the Silurian System; on the other hand, from the fact that the Upper Silurian lies unconformably on the Upper Ordovician, it may be inferred that
there might be a local tectonic movement between the Ordovician and Silurian systems. From the terminal Silurian to Middle Devonian, there occurred a strong movement which brought about folding of the Early Paleozoic
geosynclinal sediments in Borohore in the northern Tianshan Mts., so as to
form linear folds overturning to the south. Accompanying this movement,
small granite masses were formed in the Borohoro Mt. area and on the northern margin of the southern Tianshan. Glaucophane-schist zones of Tekes
in the southern Tianshan might be the products of this movement, which resulted in strong differentiation of the geosyncline; and thus the southern
and northern Tianshan Geosynclines had developed in two different directions. In the southern Tianshan the eugeosynclinal setting became a miogeosynclinal one gradually; while in the northern Tianshan, the case is reversed, i.e. the miogeosynclinal setting was turned into an eugeosynclinal one.
Not only the southern and northern Tianshan Geosynclines had been distinctly differentiated, but also the eastern and western segments of the northern Tianshan had also been subjected to distinct differentiation. In the western segment of the northern Tianshan, the folded sediments in Borohoro
outcrop as an island chain. To the north of it, there occur Middle Devonian
continental sediments and moderately thick littoral-neritic and Late Devonian paralic sediments corresponding to the cover of the platform. To the
south of it, the Lower and Middle Devonian may be absent and the Upper
Devonian may probably correspond to the cover of the platform. In the western segment, only by the Early Carboniferous did the platform sediments
begin to be once more turned into the geosynclinal sediments. It should be
pointed out that in the Carboniferous the southern side of the Caledonian
fold belt of the Borohoro Mt. was of eugeosynclinal nature, while the northern side of the fold belt was of miogeosynclinal nature, as they were separated by the Caledonian fold belt in the Borohoro Mt. In the eastern segment
of the northern Tianshan, however, the Devonian System is represented by
an over 10,000-m-thick sequence of dominantly intermediate volcanic rocks
and volcaniclastic rocks over a vast area, except in the region north of Barkol Lake where in the Early Devonian littoral sandstone and conglomerate
occurred because of the influence of the Huangcaopo Rise. The Carboniferous eugeosynclinal sediments are likewise widespread and very thick, and it

was not until the Early Permian that the scope of their distribution was reduced, the thickness became smaller and the volcanic activity weakened.

The polycyclic tectonic movements in the Tianshan Geosyncline are usually prominent. Therefore, the differentiation and migration of the geosyncline are also very apparent, and corresponding magmatic activity and mineralization are both very striking. The geological workers in Xinjiang (the Composition Group of the Xinjiang Geological Bureau, 1978) maintained that in the process of the development of the Northern Tianshan Eugeosyncline there had occurred altogether seven depositional cycles of molasse formation and basic, intermediate and acid volcanic rocks, and that under transitional conditions there had occurred several depositional cycles of carbonate, graywacke and flysch formations. We have divided the Variscan Cycle into four phases. In the early phase the eruption of intermediate and basic magmas predominated, in the middle phase the eruption of intermediate magma was dominant, in the late phase the eruption of intermediate and acid magmas and the intrusion of a large quantity of granite were prominent, and in the terminal phase many small and more alkaline rock masses occurred. The endogenic mineralization related to magmatic activity also shows a polycyclic nature. In the early phase iron mineralization was dominant; in the middle and late phases iron and polymetallic mineralizations prevailed, these three phases all being also the most important mineralization phases; in the terminal phase REE mineralizations were dominant. Mineralizations tend to differ from each other with the nature of geosynclines. In the Northern Tianshan Eugeosyncline endogenic mineralizations are dominant, while in the Southern Tianshan Miogeosyncline exogenic mineralizations are more important. It must be emphasized that the locations of magmatic activity and mineralization are changeable with structural migration, i.e. there are migrations of both magmatic activity and of mineralization.

The structural migration in the Tianshan Geosynclinal System is laterally manifested by gradual northward and southward migration of the subsidence centre, with the Tianshan Median Uplift Belt serving as a starting point of the migration (see the ensuing context). For example, in the northern Tianshan the Qoro Tagh is near the median uplift belt, while northwards, Mount Bogda is far from the median uplift belt. The geological workers of Xinjiang have pointed out (the Composition Group of the Xinjiang Geological Bureau, 1978): "From a correlation of the thicknesses of various series of the Carboniferous System between Mount Bogda and Qoro Tagh to its south, it can be seen that the thickness of deposition in Mount Bogda was only 1/12 that of Qoro Tagh in the Early Carboniferous, that the thickness of deposition in both areas was largely equal in the Middle Carboniferous and that the thickness of deposition in the former was eight times that in the latter in the late Carboniferous. It follows thus that the subsidence centre completed a lateral migration in a short period of time". The longitudinal migration shows different directions, as the migration took place in different phases and the eugeosyncline and miogeosyncline migrated in different directions. In the course of development of the geosyncline, the longitudinal migration of the Northern Tianshan Eugeosyncline and the Southern Tianshan Miogeosyncline showed reverse directions. The miogeosyncline of the preceding cycles and eugeosyncline of the principal cycle in the northern Tianshan were expressed by eastward structural migration. The geological workers of Xinjiang have also pointed this (the Composition Group of the Xinjiang Geological Bureau, 1978). The southern Tianshan Miogeosyncline was expressed by westward structural migration. For example, the Kazil Tagh area east of Bosten Lake rose after the middle Variscan Cycle; thus Middle Carboniferous or older marine sediments are absent. The Khariktau area west of Bosten Lake is marked by folds of the Late Variscan Cycle, and the Akqi area by folds of the terminal Variscan Cycle.

After the Tianshan Geosynclines were transformed into fold belts, there were developed the Kuqa Pedmont Depression to their south, the Urumqi Piedmont Depression to their north, the Turpan-Hami Intermontane Depression to their east and the Ili Intermontane Depression to their west. From these depressions, the trend of structural migration during the Mesozoic and Cenozoic may be known: longitudinally it was expressed by westward migration, which indicates that the fault blocks of succeeding cycles uplifted from east to west.

The Tianshan Geosynclinal Fold System may be divided into four fold belts: the Southern Tianshan Miogeosynclinal Fold Belt, the Northern Tianshan Eugeosynclinal Fold Belt, the Tianshan Median Uplift Belt and the Beishan Fold Belt. Of these, the first two fold belts show features such as mentioned above. The Tianshan Median Uplift Belt is composed mainly of highly metamorphosed gneiss and schist, with greenschist occurring in some places. There are three different views on the age of its formation: (1) Precambrian, (2) Early Paleozoic, and (3) Late Paleozoic. We tentatively treat the strata west of Kumux as being of Early Paleozoic age, but cannot rule out the possibility that there are Precambrian strata outcropping as an island chain. The Early Carboniferous sequences are distributed sporadically on the uplift belt. In the eastern segment of the uplift belt, blueschist may be found in the belt and its vicinity. Lying in the eastern part of the Tianshan Mts, the Beishan eugeosynclinal fold belt outcrops as low hills. The Lower Paleozoic witnessed prominent eruption of intermediate and basic magmas and the formation of a eugeosyncline. The late Caledonian Movement was strong, thus causing the Early Paleozoic strata to be tightly folded and relatively metamorphosed. In the Late Paleozoic the eruption of intermediate magmas prediominated, but in some places there is also much spilite. At the end of the Early Permian folding took place, thus bringing the geosynclinal deposition to an end.

III.7.4 The Inner Mongolian-Greater Hinggan Fold System

The Inner Mongolian-Greater Hinggan Geosynclinal Fold System occupies the central southern part of the Greater Hinggan Mts. and the Inner Mongolian grassland. Its western part also includes the area north of Alxa.

Most parts of the region are connected with the corresponding fold belts in the People's Republic of Mongolia.

The Variscan Cycle is the principal cycle of the Inner Mongolian-Greater Hinggan Fold System, and the Caledonian Cycle is its preceding cycle. In the region the main outcrops are of Late Paleozoic strata, generally not metamorphosed or very slightly metamorphosed. The Early Paleozoic strata are exposed in part of the region, generally represented by greenschist and, in a few cases, by gneiss. The tectonic lines are oriented in a nearly E-W direction to the south of the Xar Moron Deep Fracture and occur as nearly NE-oriented arcs to the north of the deep fracture.

The Early Paleozoic strata are more or less discontinuously exposed along four zones, from north to south the Hailar-Huma Zone, the Yirs-Aihui Zone, the Sonid Left Banner-Xilin Hot Zone, and the Ondor Miao-Chifeng (on the northern margin of the Inner Mongolian Axis) Zone. Along the four zones, four uplift belts were formed. On the uplift belts, the Early Paleozoic strata are unconformably overlain by the Late Paleozoic strata. Far from the uplift belts, conformable contacts may be observed between the Upper and Lower Paleozoic in a few places. So these uplift belts outcrop as island chains in the Variscan ocean. After the island chains formed in the Caledonian, the geosynclines continued to develop in the Variscan. North of the Yirs-Aihui line, geosynclinal deposition ended somewhat earlier, and since the deposition of the Early Carboniferous Morgen River Formation geosynclinal deposition has no longer occurred. So in the map the geosynclinal folds are assigned to the middle Variscides. Afterwards, the geosynclines migrated southwards up to the south of the Yirs-Aihui Zone, in which the Middle and Late Carboniferous and Early Permian geosynclinal clastic rocks and intermediate and acid volcanic rocks continued to develop. The geosynclines were folded by the end of the Early Permian. South of the Xar Moron Deep Fracture, geosynclines developed again during the Variscan Orogeny after the Caledonian folding, but according to the Stratigraphic Table of Inner Mongolia, there occurred in the main the micgeosynclinal deposits.

From the above it may be roughly construed that the northern and southern flanks of the Inner Mongolian-Greater Hinggan Geosynclinal Fold System were folded earlier, and that the folding inwards from both flanks occurred later. The geosyncline concluded its development at the end of the Early Permian.

Large amounts of ultrabasic rocks are exposed in the Solon Mt., Eren and the Sonid Left Banner as well as in the areas east of them in Inner Mongolia, and there occur submarine basic extrusive rocks and iron-bearing jasper rocks in different regions. Most geologists assign these ultrabasic rocks to the Variscan, but the authors infer, based on the intimate relations between some ultrabasic rocks and Early Paleozoic metamorphic rocks, that part of the ultrabasic rocks should be assigned to the Caledonian. Also, the authors have considered that most of these ultrabasic rocks may probably be ophiolites. In the light of field data, Hu Rao et al. (1979) have confirmed this inference.

According to Hu Rao et al., there are also glaucophane-schists and mélanges in the Sonid Left Banner.

Based on the differences in formations (Wu Changsheng 1979) and biotas (Su Yangzheng, Piao Kuangao, Zhan Lipei, pers. commun.) between the northern and southern sides of the Xar Moron Deep Fracture, the authers hold that this fracture must represent a suture of plates, that is, the final closure of the Inner Mongolian-Greater Hinggan Geosyncline took place in the Xar Moron-Mount Solon area. But Liu Changan et al. persist in its final closure along the Sain Sanda line in Mongolia (Liu Changan and Shan Ficai 1979).

The Inner Mongolian-Greater Hinggan Fold System may be further divided into the Inner Mongolian Eugeosynclinal Fold Belt and the Greater Hinggan Eugeosynclinal Fold Belt. Approximately, with the Yirs-Aihui Zone as boundary, south of the boundary lies the Inner Mongolian belt; and north of the boundary, the Greater Hinggan belt. The Greater Hinggan Mountains belong to the middle Variscides, that is, the geosynclinal deposition terminated at the end of the Early Carboniferous; while the Carboniferous-Permian sediments on them belong to the cover. Inner Mongolia belongs to the terminal Variscides, that is, the geosynclinal deposition concluded at the end of the Early Permian; while the Late Permian sediments on them belong to the cover. The Inner Mongolian belt used to be a segment where the Central Asian-Mongolian ocean was finally closed, and so is of major tectonic significance.

III.7.5 The Jilin-Heilongjiang Fold System

The Jilin-Heilongjiang Fold System is located in the Zhangguangcai Mt. and the Lesser Hinggan Mts. It is bounded on the south by the Jiaoliao Platform Uplift of the Sino-Korean Platform, on the west by the Inner Mongolian-Greater Hinggan Fold System, and on the east by the Nadanhada Fold Belt.

The Jilin-Heilongjiang Fold System is composed of several tectonic units of different nature. From east to west they are the Jiamusi Uplift, the Zhangguangcailing Eugeosynclinal Belt, the Songliao Depression, as well as the Aihui Uplift in the north. The Tongjiang Sag east of Jiamusi rests partly on the basement of the Jiamusi Uplift and partly on the basement of the Yanshanian Fold Belt at Nadanhada. Stretching in a nearly N-S direction, the Jiamusi Uplift separates the Tianshan-Hinggan Geosynclinal Region from the Marginal-Pacific Fold Region in East Asia, and itself is a single unit at the easternmost end of the Tianshan-Hinggan region. This uplift is composed of gneiss, schist and marble, termed the Mashan and Heilongjiang groups respectively. Previously, they were thought to be of Archean and Early Proterozoic age. But in the process of data collection and map compilation, we found that this metamorphic series can be correlated with the Upper Proterozoic to Lower Cambrian in the adjacent Soviet Union both in lithology and stratigraphic sequence. In accordance with the above-mentioned fact, and in combination with the tectonic setting and the isotopic ages (600—900 my), Ren Jishun and Zhang Qinwen inferred that this series should be assigned to late Proterozoic to Early Cambrian in age, and locally may be even older. In 1975, Zhang Qinwen, Xu Zhiqin et al. (unpublished data) observed and studied this rock series and discovered suspicious fossil remains, which is attactive to the interested workers. In 1977, the No. 1 Geological Party of Heilongjiang Province first discovered fossils from this rock series, which can be used for age assignment. These fossils were identified by Liu Xiaoliang (1978) from the Shengyang Institute of Geology and Mineral Resources, Chinese Academy of Geological Sciences, as *Rangea* sp., *Pteridinium* sp., etc., resembling the Ediacara fauna, whose age is thought to be Sinian.

The Jiamusi Uplift belongs to the Xingkaiides. After the geosyncline was transformed into a fold belt at the end of the Early Cambrian, the area was in a state of uplifting in the Early Paleozoic. In the Late Paleozoic, slight transgression took place, and clastic rocks and small amounts of carbonate rocks were deposited in some places, but the eastern and western parts were still separated apart from each other, displaying different environments. After further uplifting in the Paleozoic, the area remained in a state of denudation for a long period.

The Zhangguangcailing Eugeosynclinal Fold Belt is bounded on the east by the Jiamusi Uplift along the Mudan River Deep Fracture, and is characterized by wide distribution of Variscan granites on the surface. In the light of the strata exposed sporadically, it may be judged that this belt belongs to the terminal Variscan Fold Belt. The Lower Paleozoic consists of metamorphosed shale and sandstone with volcanic rocks, and the Upper Paleozoic comprises very thick sandstone, slate and limestone and intermediate and acid lavas with their clastic equivalents. Based on the differences in lithofacies and metamorphic grade, it is inferred that there might be a tectonic movement between the Upper and Lower Paleozoic. Because of scattered exposures and lack of data, the paleotectonic setting is difficult to clarify for the time being. There the Carboniferous sequence has a considerable thickness and contains plant fossils.

It should be pointed out that the Jilin-Heilongjiang Geosynclinal System might not be a unified geosyncline in the Late Paleozoic. Su Yangzheng informed us personally that in the Permian there occurred warm-water fusulinid fossils to the south of the Hulan-Mudanjiang line, while they were absent to the north of the line. This is a problem that deserves our attention.

The Songliao Depression is of Late Mesozoic age, in which a considerably thick Cretaceous oil-bearing rock series was deposited, thus giving rise to the Daqing oil field, the largest oil field in China. There are two different views about the nature of the basement of the Songliao Depression: One considers it as being a crystalline basement of the ancient rock series, and the other advocates that it should be a basement of the Variscan folds. The present authors tend to consider that there is an elongated fault block of the Xingkaiides in the depression, which trends NE and may probably be connected with the Aihui Rise. The metamorphic rocks encountered in drilling and the geophysical anomalies may be the manifestations of the basement block.

III.8 THE KUNLUN-QINLING GEOSYNCLINAL FOLD REGION

The Kunlun-Qinling Geosynclinal Fold Region is composed of the Qilian, Western Kunlun, Eastern Kunlun, and Qinling Fold Systems. It is bounded by the Tarim Platform and the Sino-Korean Paraplatform on the north, the Yangtze Paraplatform on the southeast, and the Yunnan-Tibet Geosynclinal Fold Region on the southwest. The boundaries are all represented by deep fractures.

According to the data available, we preliminarily consider that most parts of the Kunlun-Qinling Geosynclinal Region might be a geosynclinal region that was formed by taphrogeny of the Chinese Protoplatform (Huang Jiqing et al. 1977). The geosyncline began to develop largely in the Middle Cambrian, and was again gradually transformed into a fold region through its development in the Caledonian and Variscan Cycles. This process terminated at the end of the Early Permian. Therefore, it must be a Paleozoic Geosynclinal Fold Region. But in the Indosinian Cycle the Qinling Geosynclinal System continued its geosynclinal development, which was transformed into a fold belt only in Middle and Late Triassic times. In fact, by that time the Indosinian Qinling Geosyncline had already become a part of the Tethyan Yunnan-Tibet Geosynclinal Region.

III.8.1 The Qilian Fold System

The Qilian Fold System, NW-trending in general, occupies the area where the Qilian Mts. lie. This is a Caledonian fold system. On the south it is bounded by the Kunlun Fold System along the Nanshan Deep Fracture in Danghe, and by the Qinling Fold System along the Tianshui-Baoji Deep Fracture and the Nanshan Deep Fracture in Qinghai (the western segment of the Qinghai Lake-Northern Huaiyang Fracture). On the north it shows a transitional relation to the Sino-Korean Paraplatform, with the northwestern end being cut by the Altun Deep Fracture.

The Qilian Geosynclinal System is obviously divided into four second-order units that show their own distinguishing features. They are: the Corridor Transitional Belt, the Northern Qilian Eugeosynclinal Fold Belt, the Qilian Median Uplift Belt, and the Southern Qilian Fold Belt.

1. *The Northern Qilian Eugeosynclinal Fold Belt*

This has been proved to be a well-developed eugeosynclinal belt (Wang Quan et al. 1976; Xiao Xuchang et al. 1978). The inner part of the geosyncline has been differentiated greatly. Against the background of a deep depression there are several uplift belts composed of Late Proterozoic Yangtze Fold Belts. In the deep depression belt are developed the Middle and Late Cambrian and Early and Middle Ordovician ophiolites composed of ultrabasic rocks, basic extrusive rocks (spilite) with pillow structure, radiolarian siliceous rocks and pelagic flysch, more than 10,000 m in thickness. At the end of the Middle Ordovician, the Northern Qilian Geosyncline had been subjected to the strong early Caledonian Movement, which caused the Upper Ordovician to lie unconformably on the above-mentioned ophiolites and the geosyncline itself to be differentiated. The Upper Ordovician consists of graptolitic shale, carbonate rocks and andesitic volcanic rocks. The Silurian period was on the eve of the closing of the geosyncline. In this period the rocks were mainly represented by flysch and flysch-like clastic sediments. But to the north of Tianzhu there occurs a deep depression belt, where there are indications of strong volcanic activity and an andesitic volcanic rock series with a thickness of 3000 m. By the Middle and Late Silurian the seawater withdrew gradually. From the bottom upwards, the sediments become coarser, consisting mainly of clastic rocks with limestone, as well as of red sediments and conglomerate.

The Corridor Transitional Belt lies between the Northern Qilian Eugeosynclinal Fold Belt and the Alax Platform Uplift of the Sino-Korean Paraplatform. This is a continental shelf-type miogeosynclinal belt. The Middle and Upper Cambrian and Middle and Lower Ordovician are represented by clastic and carbonate sediments, nearly 10,000 m thick. In the Early and Middle Ordovician, large-scale subsidence took place near the northern Qilian Eugeosynclinal Belt, accompanied by volcanic eruptions of intermediate and basic magmas. The Upper Ordovician, overlying the Middle Ordovician unconformably, consists mainly of graptolitic shale and coquina, and the Silurian System is also represented by flysch and flysch-like sediments. From south to north the thickness and lithofacies of the entire Paleozoic sediments obviously show the features of transition from a eugeosynclinal type to a platform type.

2. The Southern Qilian Geosynclinal Fold Belt

At present the degree of study of this belt is still low. Preliminary data indicate that it is also a eugeosynclinal belt, but appears to be different from the Northern Qilian Eugeosynclinal Belt. Except in the Laji Mountain area where an ophiolite suite is well developed, the volcanic rocks of Middle and Late Cambrian to Middle and Early Ordovician are generally andesitic, andesite-basaltic and dacitic. Big ultrabasic-basic masses are only observed at Muli and other places along the deep fracture zone on the margin of the central Qilian Mts. The Late Ordovician to Silurian strata there are similar to those in the northern Qilian Mts.

3. The Qilian Median Uplift Belt

This is a remnant block of the folded basement of the Chinese Protoplatform, lying in between the Southern and Northern Qilian eugeosynclinal belts, with a deep fracture zone separating it from both the Northern and Southern Qilian belts. It is composed of metamorphic complexes of the Yangtzeides formed at the end of the Late Proterozoic. In the Early Paleozoic it was in a state of uplifting. Early Paleozoic sandstone, slate, marlstone and red conglomerate are exposed only along the fracture zone west of Mount Liuhuang in the western part of the belt; they are not very thick, but contain Ordovician fossils.

The late Caledonian Movement of the Late Silurian swept the entire Qilian Geosynclinal System, giving rise to a mainly NW-trending linear tight fold system and gigantic compressive fractures. Acid masses composed essentially of granites are mainly distributed in the Qilian Median Uplift Belt, on the northern side of the Northern Qilian Mts. and on the southern margin of Alxa. At the pediment of the Northern Qilian Fold Belt, very thick Devonian red molasse lies unconformably at a high angle on pre-Devonian metamorphic rocks and some of the above-mentioned granite masses.

The Qilian Geosynclinal System was transformed into a fold system through the late Caledonian Movement and was linked with the Sino-Korean Paraplatform to its north. Afterwards, it experienced a strong differential uplift and peneplanation through the whole Devonian period. In Carboniferous times it was again covered by seawater, and thus neritic limestone and paralic coal-bearing rock series were deposited. In Permian and Triassic times, with the deep fracture on the northern margin of the central Qilian Mts. as a boundary, to the north of it were deposited the continental basin sediments, while neritic-littoral sediments were deposited to the south of it. After the Late Triassic Indosinian Movement, with the formation of the gigantic Indosinian Fold System in southwestern China, the whole region where the Qilian Fold System lies was also uplifted and became a land, thus bringing the history of marine transgression to an end.

III. 8.2 The Qinling Fold System[6]

The Qinling Fold System is located in the heart of China. On the north, it adjoins the Sino-Korean Paraplatform along the deep fracture on the northern margin of the Qinling Axis (its eastern extension is known as the Queshan-Feidong Deep Fracture) and is separated from the Qilian Fold System by the Tianshui-Baoji Fracture; to the south, it is bounded by the Yangtze Paraplatform along the Chengkou-Fangxian Deep Fracture and the Xiangfan-Guangji Deep Fracture, and by the Songpan-Garzê Fold System along the Maqen-Lueyang Deep Fracture; its eastern end is cut by the Tancheng-Lujiang Deep Fracture; its western end is inserted between the Kunlun Fold System and the Qilian Fold System south of the Qinghai Lake.

The Qinling Fold System is a typical polycyclic fold system. It may be further divided into five second-order units, i.e. the Northern Qinling Ca-

---

6 Mainly according to the unpublished data of the Regional Survey Brigades of Shaanxi. Gansu, etc.

ledonian Geosynclinal Fold Belt, the Lixian-Zhashui Variscan Fold Belt, the southern Qinling Indosinian Miogeosynclinal Fold Belt, the Northern Daba Mt. Fold Belt and the Wudang-Huaiyang Uplift.

1. *The Northern Qinling Caledonian Fold Belt*

It is actually the eastern extension of the Qilian Caledonian Fold System, but as it had been further subjected to a strong modification by subsequent polycyclic tectono-magmatism, people have for quite a long time regarded it as the southern margin of the Sino-Korean Paraplatform and called it the Qinling Axis.

The Northern Qinling Caledonian Fold Belt may be subdivided into three subbelts:

The northern subbelt, i.e. the Iushi-Luanchuan area, is a miogeosynclinal belt, composed mainly of Cambrian rocks. To the north, it seems to show a transitional relationship with the Sino-Korean Paraplatform.

The central subbelt, i.e. the median uplift belt, embraces the northern half of the Qinling Axis, composed of the Late Proterozoic Kuanping and Taowan Groups. As the Kuanping Group contains spore fossils that occur in the middle and upper parts of the Late Proterozoic sequence from Jixian, North China, it was assigned to the Yangtze Fold Belt by Huang Jiqing et al. in 1974. Recent studies have fully confirmed this argument.

The southern subbelt, i.e. the southern eugeosynclinal fold belt, was originally considered to be the main part of the Qinling Axis. As Cambrian-Ordovician brachiopods and microflora have been found from it, most metamorphic ages of the rocks fall within 420 my, and the earliest phase of granites intruded into it ranges in age from 380 to 420 my, it is now assigned to the Caledonian Fold Belt. But the possibility that the Proterozoic strata outcrop as island chains is not ruled out here.

2. *The Lixian-Zhashui Variscan Miogeosynclinal Fold Belt*

This is bounded on the south by the Southern Qinling Indosinian Miogeosynclinal Fold Belt along the Lintan-Shanyang Deep Fracture. Its western segment (west of Fengxian) is marked by a very thick flysch formation of the Middle Devonian Xihanshui Group, which is overlain unconformably by molasses of the Upper Devonian Dacaotan Group; in the middle segment (south of Baoji), the Middle Carboniferous Caoliangyi coal measures lie unconformably on the Devonian-Lower Carboniferous flysch and carbonate rocks; in the eastern segment (the Zhashui area), the Middle Carboniferous (possibly the Upper Carboniferous included) remains geosynclinal sediments, and this segment had not been transformed into a Variscan fold belt until the Permian

3. *The Southern Qinling Indosinian Miogeosynclinal Fold Belt*

This is located between the Lintan-Shanyang Deep Fracture and the Narqen-Lueyang and Shiquan-Ankang Deep Fractures, and extends westwards through the Nanshan Mts. in Qinghai Province and inserts between the Kunlun Fold System and the Qilian Fold System, while its eastern part overlaps the Wudang-Huaiyang Uplift.

In this fold belt, the Sinian to Ordovisian strata are essentially of the platform type, and geosynclinal deposition did not begin until the Silurian. In the western segment, the oldest strata exposed in the Tewo-Wudu area are Silurian intermediate-basic metavolcanic rocks, showing that the Caledonian Cycle is furnished with eugeosynclinal features. The lower part of the Silurian System consists of graptolitic shale, intercalated with carbonaceous and siliceous rocks, whereas its upper part is composed of sandy shale and limestone, intercalated with carbonaceous and siliceous rocks. The Devonian to Lower Permian strata are almost all composed of carbonate rocks. The Triassic System is mainly distributed in Fengxian County and to its west, all being represented by the flysch formation. East of Hezuo only the Middle and Lower Triassic have been found, while west of it the Upper Triassic has been observed as well and there occur many volcanic rocks.

It is worthwhile to note that in this geosynclinal belt the Sinian to Triassic strata are in the main continuous and that local unconformities only occur in between the Silurian and the Devonian systems to the east of Yuguan (Shaanxi Regional Survey Party 1968). Important tectonic movements took place in the Middle and Late Triassic times. The Late Triassic continental coal-bearing volcanic molasses are observed to rest unconformably on the Middle and Early Triassic geosynclinal sediments. Thus it is a typical Indosinian miogeosynclinal fold belt.

What ought to be explained is that the digitate Indosinian geosynclinal fold belt which is inserted between the Kunlun and Qilian Fold Systems in an area extending from the Nanshan Mts. to the Central Wunong Mt. of Qinghai Province represents a very peculiar segment of the Qinling Indosinian Geosyncline. This segment was originally a part of the Eastern Kunlun Variscan Fold System. There the Permian System consists of less thicker platform sediments, while the Triassic System consists of very thick, well-developed flysch formations and is connected with that of the Qinling Indosinian Geosyncline. The Jurassic coal measures lie with high-angle unconformities on Triassic geosynclinal sediments.

4. The Northern Daba Mt. Caledonian Fold Belt

Lying south of the Shiquan-Ankang Deep Fracture, this bounds the Yangtze Paraplatform along the Chengkou-Fangxian Deep Fracture. It is mainly composed of Cambrian-Silurian sediments. Its lower part is composed of sandy slate, intercalated with carbonate rocks, while its upper part consists mainly of sandy slate with abundant stone coals. The appearance of a large quantity of diabase dyke swarms is a very important feature for this fold belt.

The Northern Daba Mt. Fold Belt has long been considered to be a Caledonian fold belt; but as Devonian-Triassic sediments are absent above the

Early Paleozoic geosynclinal sediments and, even more, as there are no unconformities between the Devonian and the Silurian systems. In recent years some have doubted its assignment to the Caledonian Fold Belt and consider it to be an uplift that has not been subject to Caledonian folding.

5. *The Wudang-Huaiyang Uplift*

The problem concerning the "Huiyang Shield" and its assignment is one of the major problems that remain to be solved in the study of the geotectonics of China. Now it has been ascertained preliminarily that its core part is represented by the highly metamorphosed Dabie Group, with an isotopic age of 2080 my (Chen Ziqiang 1979). Its main part is the Late Proterozoic Hongan Group and its equivalents, belonging to the Yangtze Fold Belt. Considering that it has been structurally intimately related to the Qinling Fold System since the Paleozoic and constitutes a very active geoanticlinal uplift belt that plunges to the west and tilts up to the east in the Qinling Fold System, so we tentatively assign it to the Qinling Fold System.

III.8.3 The Eastern Kunlun Fold System

Here we take the Ruoqiang-Lazhulong fault as a boundary to separate the Kunlun Fold System into two parts: the western Kunlun and eastern Kunlun Fold Systems mainly according to the data of regional surveys in Qinghai Province.

The northern part of the eastern Kunlun Fold System adjoins the Qilian Fold System; its southern part bounds the Songpan-Garzê Fold System along the deep fracture on the southern margin of the eastern Kunlun Mts.; its western part is separated from the Tarim Platform by the Ruoqiang-Lazhulong Deep Fracture; and its easternmost part is "overlapped" by the Qinling Indosinian Geosynclinal Fold Belt in the Xinghai area east of Dulan.

The eastern Kunlun Fold System is a complex Variscan fold system, and may be divided into seven second-order units, i.e. the Dakendaban Eugeosynclinal Fold Belt, the Oulongbuluk Uplift Belt, the eugeosynclinal fold belt on the northern margin of the Qaidam Basin, the Qaidam Depression, the Chimantagh Eugeosynclinal Fold Belt, the Burhan Budai Eugeosynclinal Fold Belt, and the Altun Eugeosynclinal Fold Belt.

As is the case with the Qilian Geosyncline, the Eastern Kunlun Fold System was also formed by disintegration of the Chinese Protoplatform. The remnant fragments of the Yangtze Fold in the geosyncline are mainly exposed in the Oulongbuluk Uplift Belt, the Eastern Kunlun Median Uplift Belt and the Altun Fold Belt. In recent years, Sinian platform strata (conglomerate and sandstone in the lower part; quartz sandstone and siliceous rocks with iron formations in the middle part; and carbonate rocks in the upper part; with tillite in some places; 1000 m or so in thickness; in the main not being metamorphosed), have been found one after another in Oulongbuluk, Quanji and Urt Moron (north of the eastern Kunlun Median Uplift Belt) as well as at Annanba of Altun Mts. They lie unconformably on the Upper

Proterozoic or older rocks that were subject to strong folding and regional metamorphism. In Burhan Budai, the lower part of the original rocks of the Upper Proterozoic strata was composed mainly of terrigenous clastic rocks, intercalated with intermediate-basic volcanic rocks; and the upper part was composed of carbonate rocks, intercalated with pelitic shale, containing abundant stromatolites, and yielding a great thickness. After having been subject to intermediate-grade regional metamorphism, their lower part has become amphibolite facies and the upper part is of greenschist facies with local migmatization, whose age is roughly equivalent to that of the Late Proterozoic sequence with its standard section in Jixian County (unpublished data from the Qinghai Regional Survey Party and Qu Jingchuan from our division). The Cambrian to Devonian strata of the eastern Kunlun Mts. include two different associations of sedimentary formations.

The platform formation association consists mainly of carbonate rocks, being only observed in the Oulongbuluk Uplift Belt. The Cambrian to Early Ordovician strata consist of limestone with small amounts of shale; the Middle Ordovician is represented by graptolitic shale. The entire association is 2000 m thick. The Late Ordovician to Middle Devonian strata are absent.

The relatively typical ophiolite suite in the geosynclinal formation association is only developed in the eugeosynclinal fold belt on the northern margin of the Qaidam Basin (which extends eastwards up to the southwest of the Caka Salt Lake). In other areas there mainly occur several eruptive-sedimentary cycles formed by intermediate-basic and intermediate-acid volcanic rocks, volcaniclastic rocks, sedimentary clastic rocks and carbonate rocks, which have been turned into greenschist, phyllite and slate through slight regional metamorphism. Qin Deyu, Li Guangcen et al. (1974, unpublished data) found abundant fossils of cephalopods, brachiopods, corals and gastropods from this volcano-sedimentary series when they conducted a geological investigation in an area from Golmud to the Kunlun Pass in 1974. These fossils include *Maclurites* sp., *Poleumita* sp., *Favistella alveotala*, *Plasmoporella chingueiebsi*, *Wormspora* sp., *Mesofavosites* sp., *Enryspirifor* sp. and *Sinospirifer* sp. (identified by Lin Baoyu et al. from our Institute, unpublished data), showing an age of Ordovician to Devonian. As there is still a very thick sequence below the fossiliferous strata, the lower part of the series might also include the Cambrian System.

The early Variscan Movement that took place in the Late Devonian is of great significance to the development of the Eastern Kunlun Geosyncline. On the northern margin of the Qaidam Basin and in the Burhan Budai area, the above mentioned Ordovician-Devonian slightly metamorphosed series is observed to be overlain unconformably by very thick Late Devonian, intermediate-basic and intermediate-acid volcanic rocks containing fossil fragments of brachiopods and plant fossils of *Lepidodendron,* which are in turn overlain unconformably by Late Devonian-Early Carboniferous red molasses. Wang Zewen considers it to be of Early Carboniferous. This indicates that by that time the development of the geosyncline had ended. In the Carboniferous to Permian, sediments had been transformed into carbonate formations of platform type.

It must be pointed out, however, that the Altun and Qimantagh Eugeosynclinal Belts did not conclude their development, although they were also affected by the Early Variscan Movement. In the Jinyan Mt. area on the northern margin of the Altun Mts. the Upper Devonian to Lower Carboniferous sequences consist of very thick andesitic volcanic rocks and flysch-like sediments. The closure of the geosyncline occurred at the end of the Early Carboniferous. The Middle and Upper Carboniferous sequences are composed of carbonate formations of platform type. In Qimantagh, the sequences of Devonian to Lower Permian are marked by very thick intermediate-basis volcanic rocks, intercalated with carbonate rocks and flysch-like sediments. Near the southern side of the Altun Deep Fracture there also occur ultrabasic rocks which are unconformably overlain by a Triassic red molasse formation. This indicates that the development of the geosyncline did not conclude until the end of the Paleozoic Era. In the Dakendaban Eugeosynclinal Fold Belt, the geosynclinal deposition did not stop until the end of the Carboniferous, the sediments of which are unconformably overlain by only a few hundreds of meters of Permian sediments of platform type. These facts tend to show that the Variscan Movement occupies a quite important position in the development of the Kunlun Geosynclinal System. This argument may also be evidenced by the gigantic Burhan Budai granite belt (with isotopic ages ranging from 260—280 my) and the granite belt on the northern margin of the Qaidam Basin formed in the Late Variscan stage.

Besides, some signs indicate that the early Caledonian Movement taking place between the Middle and Late Ordovician is also of certain significance to the development of the Eastern Kunlun Geosyncline. In the Oulongbuluk Uplift Belt, the Late Ordovician to Devonian sediments are absent; in Burhan Budai, the Upper Ordovician lies on the underlying strata with a low-angle unconformity, and large granite gravels are contained in basal conglomerates.

The Qaidam Depression is a large Meso-Cenozoic intermontane depression. It was the result of strong activity in the Tethys-Himalayan Tectonic Domain in Western China since the Mesozoic, particularly the Cenozoic. The Qaidam Depression used to be considered as a median massif, but now we have in the main negated the existence of this massif on the basis of the fact that the northern margin of the Qaidam Basin has been determined to be an eugeosynclinal fold belt and that the Qimantagh Eugeosynclinal Belt extends eastwards into the Qaidam Basin. This relatively reasonable explanation shows that the basement of the Qaidam Basin resembles an "assorted cold dish": its northern half is marked by the early Variscan eugeosynclinal folded basement of the northern margin of Qaidam, the Mangnai area in its western segment is marked by the late Variscan folded basement of Qimantagh, and Dabuson Lake and the Qaidam River area in its eastern segment are marked by the Yangtze folded basement of the Burhan Budai Median Uplift Belt.

III.8.4 The Western Kunlun Fold System

The Western Kunlun Fold System is located to the west of the Ruo-qiang-Lazhulong Fracture. It occurs as a reversed S-shaped fold mountain system towering to the southwest of the Tarim Platform.

At present the study of the Western Kunlun Fold System is still at a low level. The data available have preliminarily shown that the fold system seems to have experienced an evolutional history roughly similar to that of the Eastern Kunlun Fold System, but there are also some very important differences. The similarities between them are as follows: the development of the geosyncline initiated at the beginning of the Paleozoic Era after the disintegration of the Chinese Protoplatform; the early Variscan Movement of the Late Devonian caused its folding for the first time; the Tisnaf molasse of the Late Devonian containing *Leptophloeum rhombicium* and *Lepidodendron* is widespread in the geosyncline; the development of the geosyncline terminated in the Early Permian.

The principal differences between it and the Eastern Kunlun Fold System are as follows

1. The Sinian System is composed of sandstone, boulder clay (tillite), submarine basic volcanic rocks and carbonate rocks, 4500 m in thickness, showing a great mobility.

2. The Carboniferous-Permian geosynclinal formations deposited after the early Variscan Movement are represented mainly by terrigenous clastic rocks and carbonate rocks, while volcanic rocks are rare. That is to say, the Upper Paleozoic shows the features characteristic of a miogeosyncline, which are different from those of the eugeosyncline in the Eastern Kunlun Mts.

The Western Kunlun Fold System may be tentatively divided into three second-order units, i.e. the Northern Kunlun Miogeosynclinal Fold Belt, the Kunlun Median Uplift Belt, and the Southern Kunlun Fold Belt.

III.9 THE YUNNAN-TIBET GEOSYNCLINAL FOLD REGION

The Yunnan-Tibet Geosynclinal Fold Region is an important component part of the gigantic Mesozoic northern Tethyan geosynclinal belt on the southern margin of Eurasia and occupies the main part of the famous Qing-hai-Tibet Plateau. It embraces a wide region to the south of the deep fracture on the southern margin of the Eastern Kunlun, to the north of the Yarlung Zangbo River Deep Fracture, and to the west of the Longment Mt. Deep Fracture and the Jinsha River-Red River Deep Fracture. With the Jinsha River-Red River Deep Fracture and the Bangong Lake-Nujiang Deep Fracture as boundaries, the fold region may be divided into three parts: the Indosinian Songpan-Garzê Geosynclinal System (north of the Jinsha River-Red River Deep Fracture), the Yanshanian Karakorum-Tanggula Geosynclinal Fold System and the Indosinian Sanjiang (Three-River) Geosynclinal Fold System (south of the Jinsha River-Red River Deep Fracture), and the

Yanshanian Gangdisê-Nyainqêntanglha Geosynclinal System (south of the Bangong Lake-Nujiang Deep Fracture).

The main characteristics of this geosynclinal fold region are: (1) the eugeosynclinal belt and the miogeosynclinal belt are arranged alternately, i.e. the eugeosynclinal belts occur along the deepefracture zones, while a miogeosynclinal belt occurs in between them; (2) several small median massifs of different ages are hidden under the Mesozoic sediments in Zoigê, Qiangtang, Baoshan, Lanping-Simao and Qamdo areas (but in Baoshan under the Paleozoic sediments); (3) the geosyncl'ne has migrated from north to south, with the Early Mesozoic Indosinian Fold System lying in the north and the Late Mesozoic Yanshanian Fold System lying in the south.

III.9.1 The Songpan-Garzê Fold System[7]

The Songpan-Garzê Fold System occupies a triangular area south of the Eastern Kunlun Deep Fracture, west of the Longmen Mt. Deep Fracture, north and east of the Jinsha River--Red River Deep Fracture. This is an Indosinian Geosynclinal Fold System. The entire area has been almost completely covered by Triassic geosynclinal sediments, the Paleozoic rocks being exposed only in the eastern marginal part adjoining the Yangtze Paraplatform.

The Songpan-Garzê Fold System may be divided into three second-order units, i.e. the Bayan Har Fold Belt, the Yarlung Zangbo Fold Belt and the Yushu-Yidun Fold Belt. Each is composed of an eugeosyncline and a miogeosyncline. The eugeosynclinal belts in the north or south are distributed along the western or southern sides of the gigantic deep fracture zones.

1. *Bayan Har Fold Belt*

The northern subbelt, i.e. the A'nyêmaqen Eugeosynclinal Fold Belt, is distributed along the southern side of he deep fracture on the southern margin of the Eastern Kunlun Mts. The geosynclinal formations consist mainly of Permian-Triassic spillitic keratophyre and flysch, with small amounts of carbonate rocks. Ultrabasic rocks and mélanges are widely distributed over a distance of 800 km from the Kunlun Pass to Maqu.

The southern subbelt, i.e. the Bayan Har Miogeosynclinal Fold Belt, occupies the main part of the Bayan Har Fold Belt, composed of Middle and Late Triassic flysch, with a thickness of at least 10,000 m or more (plate IV) (Zhang Qinwen, Ren Jishun et al. (1965, unpublished data) determined the Triassic in the Xianjin and Barkam areas to be a suite of well-developed flysh formations when they worked in western Sichuan in 1965.). The Paleozoic sequence is only distributed on its eastern margin. The sequences from Cambrian to Middle Ordovician are similar to those in the

---

7 Mainly according to a mass of unpublished data of regional surveys of Sichuan and Qinghai.

Yangtze Paraplatform. For example, the Baota Limestone typical of the Yangtze Platform has been discovered in many places at the side of the geosyncline west of the Longmen Mt. Deep Fracture. The Silurian to Lower Triassic sequences are represented by sediments of transitional type. Intensive development of the geosyncline mainly took place in the Middle to Late Triassic. The Xiaotangzi Formation of the Noric stage in the Longmen Mt. area lies unconformably on the Triassic or older strata, and the Late Triassic molasse (containing coal and volcanic rocks) is accumulated in small intermontane basins in the Aba (Ngawa) area. All these facts have proved that the geosyncline was folded at the beginning of the Noric stage of the Late Triassic.

2. *Yajiang Fold Belt*

The eastern subbelt, i.e. the Luhuo (Zhaggo)-Dawu Eugeosynclinal Fold Belt, is distributed along the southern side of the Luhuo-Kangding Deep Fracture and is composed of Permian-Triassic basic volcanic rocks, flysch and carbonate rocks. It is best exposed in the Luhuo-Dawu area.

The western subbelt, i.e. the Yajiang Miogeosynclinal Fold Belt, is mainly composed of Middle-Upper Triassic flysch, and in comparison with the Bayan Har belt, the flysch in this belt is badly developed, but is intercalated with basic volcanic rocks and large lenses of carbonate rocks. Its southern part "overlaps" the Yanyuan-Lijiang Platform-Margin Depression Zone on the western margin of the Yangtze Paraplatform, exhibiting a distinct transitional relationship.

3. *Yüshu-Yidun Fold Belt*

The eastern subbelt, i.e. the Garzê-Litang Eugeosynclinal Fold Belt, spreads along the Garzê-Litang Deep Fracture Belt. In Ordovician-Silurian times, a geosynclinal belt was formed which consists mainly of clastic rocks, intercalated with volcanic rocks; but it is the Permian-Triassic eugeosynclinal belt that manifests itself most distinctly, consisting of flysch, intermediate-basic volcanic rocks and ultrabasic rocks. Recently, clues of mélanges have been found in the Garzê area.

The western subbelt, the Daocheng Geosynclinal Fold Belt, is characterized by alternating beds of clastic rocks and limestone, intercalated with more basic volcanic rocks, containing relatively abundant fossils, showing some features characteristic of eugeosynclines. But the sequences from the Cambrian to the Permian in Zhongzan and other places are represented by carbonate rocks, containing very abundant fossils, showing the features characteristic of platform formations. This indicates that the activity of the geosyncline reached its climax also in the Triassic times.

With regard to the Zoigê Massif, some Chinese geologists believed long ago that the Songpan area in northwestern Sichuan was a median massif. But as no older metamorphic basement outcrops at the surface there, there

has long been disagreement on this point of view. In the light of field profile observations, Guo Yongling again posed this problem in 1963. He called
this massif the Zoigê Massif, but this still did not arouse attention. In 1974
aeromagnetic surveys discovered that the Zoigê grassland in western Sichuan
was a positive magnetic field area (Jiang Nengqiang, oral account) distinct from other areas in Bayan Har; thus it has preliminarily been determined that a rigid massif may be hidden beneath the Triassic sediments
there. During the compilation of the 1:4,000,000-scale *Tectonic Map of China,*
we found that the so-called Zoigê Massif was displayed very clearly on
a Landsat photograph with its shape being very much like *Spirifer.* Thus
on the basis of the field observation by Guo Yongling, deep detection by aeromagnetic surveys and analysis and interpretation of satellite images, we
can absolutely confirm the existence of the concealed Zoigê Massif; however,
the age of its basement remains to be determined in the future work.

III.9.2 The Sanjiang (Three-River) Fold System[8]

The Sanjiang Fold System is an Indosinian geosynclinal fold system, lying in western Yunnan and the Qamdo area of Tibet east of the Nujiang River and west of the Jinsha River. In eastern Tibet the Jinsha, Lancang and
Nujiang Rivers turn suddenly to the south to form the famous Sanjiang
(Three-River) Valley. To the south it extends into the Indo-China Peninsula, and its western end is overlapped by the Tanggula Yanshanian Fold
System.

The Sanjiang Fold System has experienced a very complex developmental history. The oldest strata with fossil evidence in the system is the Gong-
vanghe Group of the Middle and Lower Cambrian (possibly including part
of the Sinian) in the Baoshan area. This group is represented by a very thick
low-grade metamorphic flysch-like formation, containing such fossils as
sponge spicules, overlain slightly unconformably by sandy shale of the
Baoshan Formation of the Upper Cambrian. In recent years, the geological workers in Yunnan have also found stromatolites of late Proterozoic age
in the Lancang Group, conodonts of possibly Cambrian age in the Wuliang-
shan Group and spores of late Proterozoic-Early Cambrian ages in the metamorphic rocks of the Ailao Mt. area. According to these fossils, and in
combination with the fact that the oldest unmetamorphosed strata in the
area are of the Upper Cambrian (e.g. the Baoshan Formation) of Ordovician, we preliminarily infer that the Ailao Mt. and Lancang River metamorphic series are likely to be the products of metamorphism of the Proterozoic
to Early and Middle Cambrian Xingkaiides. Perhaps it is these metamorphic rocks that form the basement of the Sanjiang Geosyncline.

After the Xingkai Cycle, in the Paleozoic, most of the Sanjiang area
was in a relatively stable paraplatform state, and the sediments are generally represented by neritic clastic rocks and carbonate rocks. The Early Pa-

---

8 Mainly according to the unpublished data of the Yunnan Regional Survey Brigade.

leozoic Caledonian geosynclinal belt has been documentarily proved to extend roughly along the western sides of the Jinsha River and the Ailao Mt. In Qingnidong east of Qamdo, the Middle Devonian basal conglomerate has been seen to lie unconformably on the graptolite-bearing flysch-like sandy slate of Ordovician age, (Sichuan Provincial Geological Bureau, 1974; Ning Qisheng, unpublished data) and west of Judian in northwestern Yunnan, the Devonian has been observed to rest unconformably on the Shigu schist of Early Paleozoic age, thus indicating that this belt is a Late Caledonian fold belt. Afterwards, a Late Paleozoic eugeosynclinal belt was developed along the western bank of the Jinsha River. Wang Kaiyuan (unpublished data) holds that a Carboniferous-Permian ophiolite suite occurs in the Dêqên-Weixi area. It appears that a serious study should be made to confirm whether the metamorphic series of the "Jiayuqiao Group" outcropping along the western bank of the Lancang River and the Early Paleozoic metamorphic series outcropping in the Changning-Gengma area on the western flank of the Lancang Metamorphic Belt in western Yunnan also represent an Early Paleozoic geosynclinal belt.

In the Sanjiang Fold System the tectonic framework, characterized by the alternate arrangement of eugeosynclines and miogeosynclines, has mainly been formed since the Triassic Indosinian Cycle. The deep-subsiding geosynclinal belts are largely distributed along the Jinsha River-Red River, Lancang River, and Nujiang River Deep Fracture Belts, where very thick Triassic volcano-sed'mentary formations of the geosynclinal type were deposited, with an ophiolite suite occurring in the Dêrong area on the western bank of the Jinsha River (Zhang Zhimeng, unpublished data). In Baoshan, Lanping-Simao and Qamdo areas between the geosynclines are relatively uplifted areas within individual geosynclines, which are characterized by relatively thin sediments and less tectonic activities.

The Late Triassic Indosinian Movement represents a main folding episode in the Sanjiang Geosynclinal System, which brought about strong folding of the Indosinian Geosynclines distributed along several deep fracture zones. On the other hand, structural disturbances were relatively slight in the Qamdo-Lanping-Simao area, where intense subsidence took place after the Indosinian Movement, thus giving rise to very thick gypsum salt-bearing red formations of Late Triassic-Early Tertiary ages, i.e. the famous western Yunnan Red Beds and the Qamdo Red Beds.

Previously the period of folding of the western Yunnan Red Beds was generally assigned to the Yanshanian Cycle, but recent work has proved that the topmost part of the continuous red formations in fact also includes the Paleogene (the Yunlong Formation and the Guolang Formation) and that the major tectonic movement took place in Late Eocene times. The Late Eocene-Oligocene rad molasse may be seen to lie unconformably on the folded red beds of western Yunnan to the west of Mengla and Lanping. This folding took place almost simultaneously with that of the Himalayan Geosyncline, so these folds should be regarded as belonging to the Himalayan Cycle.

Another important feature of the Sanjiang Fold System lies in the fact that the polycyclic tectonic movements, magmatism and metamorphism along the deep fracture zones were so strong that the fold system underwent structural disturbances and metamorphism to varying degrees not only in the Paleozoic and Mesozoic but also in Cenozoic times; thus forming some belts of polycyclic tectono-magmat'c metamorphic complexes. Of these the Ailao Mt., Lancang River and Gaoligong Mt. metamorphic belts are the most famous.

The Sanjiang Fold System may be divided into six second-order units: (1) the Ailao Mt. Eugeosynclinal Fold Belt (including Jinping-Mojiang); (2) the Jinsha River Eugeosynclinal Fold Belt; (3) the Lanping-Simao Depression; (4) the Uli-Angqi Fold Belt; (5) the Lancang River Fold Belt; and (6) the Baoshan Fold Belt which is actually the northernmost plunging part of the Shan State Massif.

III.9.3 The Karakorum-Tanggula Fold System[9]

The Karakorum-Tanggula Fold System is a Yanshanian fold system, with the western segment of the Jinsha River-Red River Deep Fracture serving as its northern boundary and the Bangong Lake-Nujiang River Deep Fracture as its southern boundary. The southeastern part of the fold system overlaps the Sanjiang Indosinian Fold System, and westwards it extends through the Karakorum Mts across the border line. This fold system is separated by the Ruoqiang-Lazhulong Deep Fracture into two segments: the western segment is called the Liqitang Geosynclinal Fold Belt, and the eastern segment is known as the Tanggula Geosynclinal Fold Belt.

For quite a long time there have been two different views about the structures in the Tanggula Range area among the Chinese geologists. Some consider it to be a platform with the Precambrian folds as the basement (Institute of Geology, Academia Sinica 1959), while others take it for a geosynclinal fold belt (Huang Jiqing 1945, 1960; Huang Jiqing et al. 1977). The major questions are: (1) does the Qiang tang area represent a Precambrian massif? and (2) do the Jurassic strata in the Tanggula Range represent geosynclinal sediments?

From the data available, it can be seen that a massif does exist in the Qiangtang area. The Expedition Team to Tibet discovered complicatedly deformed crystalline schists composed mainly of quartz-schist and mica-schist in the Mayigang Snow Mountains on the Qiangtang Plateau in 1976. On interpreting Landsat photographs, the authors found that these schists are largely distributed in an area of ca. 800 km from east to west and of ca. 300 km from north to south, extending from Qaluor County in the west to the Mayigang Snow Mt. in the east, and from the Karakorum Mts. in the north to Dabula in the south. Meanwhile, in northern Tibet, the Devonian to Triassic sequences are represented by not very thick shallow-marine, platform

---

9 Mainly after the results of the work by the Qinghai Regional Survey Brigade and Comprehensive Investigation Team of the Qinghai-Tibet Plateau, Academia Sinica, as well as part of workers from our division.

clastic rocks and carbonate rocks, containing abundant fossils. Fossils of *Gigantopteris* have even been found in Late Permian strata. These facts indicate that in Late Paleozoic-Triasis times, this area was in a relatively stable platform state and that its basement might be either Caledonides or late Preceambrian folds. Therefore, we consider that in the Late Paleozoic to Early Mesozoic (Triassic) the main part of northern Tibet was a relatively stable massif, called the Qiangtang Massif. The Sanjiang Indosinian Geosynclinal Fold Belt might have extended westwards along its northern and southern sides: while the Qiangtang, Qamdo and Lanping-Simao areas formed paternoster uplift belts or small median massifs in the entire Tanggula-Sanjiang Geosynclinal Belt.

Some maintain that the Jurassic sequence in the Tanggula is not geosynclinal sediments but platform-cover sediments. Qing Deyu and others have made field observations and laboratory studies (Qing Deyu and Li Guangceen, unpublished data), and on the basis of the following facts: there is a considerable thickness of deposits——the Togmeglha Group (T_3-J_1) being more than 4000 m thick, and the Yanshiping Group (J_{2-3}) being 4500—6000 m thick; (2) the middle member of the upper part of the Yanshiping Group on the northern slope of the Tanggula Range is marked by calcareous flysch sediments with distinct flysch imprints; and (3) the clastic rocks consist mainly of graywacke and polymineralogic sandstone; we should therefore hold that in general it might belong to the category of geosynclinal sediments, although not being very typical. Qing Deyu has also found such volcanic rocks that may possibly belong to the spilite-keratophyre sequence and red jasper rocks in the vicinity of the Amdo area along the Bangong Lake-Nujiang Deep Fracture. These rocks, together with ultrabasic rocks, form an ophiolite suite. This implies that during the Jurassic the geosyncline had migrated from east to west with a folding taking place in the Sanjiang Geosyncline, and that the Tanggula area had been transformed from a relatively stable area into a geosynclinal belt by that time. This geosyncline was folded in the Late Jurassic, causing the Early Cretaceous strata to lie unconformably or disconformably on the Yanshiping Group and on the ultrabasic masses that emplaced the Jurassic sequences.

There are few data concerning the Linqitang Fold Belt. It seems that it has a development history essentially similar to that of the Tanggula. The Devonian to Triassic systems are represented by shallow-water clastic rocks and carbonate rocks, with a thickness of more than 2500 m. The Jurassic sequence is mainly marked by shallow-water clastic rocks, interfalated with carbonate rocks, 4000—5000 m thick. The Jurassic and its underlying strata together form inverted tightly arranged linear folds and imbricate structure, on which lie unconformably the Cretaceous red sandy conglomerates.

III.9.4 The Gangdisê-Nyainqêntanglha Fold System[10]

This fold system embraces a wid area between the Bangong Lake-Nujiang River Deep Fracture and the Yarlung-Zangbo-River Deep Fracture, and constitutes the southernmost geosynclinal fold system of the Northern Tethyan Mesozoic Geosynclinal Region.

This is a late Yanshanian fold system that was closed in the Late Cretaceous. As the degree of the study done in the past is slight, we cannot describe its development history and structural characteristics accurately at present.

The several small pieces of migmatite, gneiss and crystalline schist exposed in the Longling area of Yunnan Province and the Nyainqêntanglha Range and the Ngari area of Tibet may represent the oldest metamorphic series in this fold system. The Ordovician to Silurian strata represent the oldest strata with fossil evidence. The Yunnan Regional Survey Party found the Ordovician to Silurian strata containing *Orthoceras* in the Longling area long ago. At the beginning of the 1970's, the Sichuan Regional Survey Party also found limestone outcrops containing Ordovician brachiopods in the Guqin area north of Zovü. Recently the Tibet Oil Prospecting Party and other institutions have also discovered limestone and graptolitic shale of Ordovician to Silurian ages on the western bank of Nam Lake and near the Baingoin Lake. In reviewing the Ordovician-Silurian sedimentary series in various places, we have found that this sedimentary series seems to show features of stable deposition. On the basis of these data, combined with the data obtained through aeromagnetic surveys, we suspect very much that the Nam Lake area is a remnant rigid massif, and that the Nyainqêntanglha metamorphic series might be its crystalline basement. The Upper Paleozoic is mainly distributed to the east of Damxung Fracture and comprises the Devonian, Carboniferous and Permian sequences, having a considerable thickness of more than 7000 m. Since the Mesozoic, the structures there have undergone strong differentiation, giving rise to the formation of the Nagqu and Lhasa Geosynclinal Fold Belts, with the Gar-Lhari Deep Fracture Zone as the boundary between them.

1. *Nagqu Geosynclinal Fold Belt*

Recent work has revealed that the main part of the proposed Sadêng Slate distributed widely in the Nagqu-Dênqên area of northern Tibet does not belong to the Triassic but Jurassic-Early Cretaceous in age (after the unpublished data of the Comprehensive Research Party of the Tibet Autonomous Region Bureau of Geology and Mineral Resources, and of the Institute of Geology, Chinese Academy of Geological Sciences). There the Triassic sequence is only exposed in the Sog-Jilutong area. It is a suite of monofacial flysch sediments of simple lithology, overlain unconformably by the Upper Triassic-Lower Jurassic coal measures. The Middle and Upper Jurassic are similar to those in the Tanggula Fold System, consisting of alternating beds

---

10 Mainly according to the data obtained through field observations by the Geological Bureaus of Yunnan Province and Tibet Autonomous Region and the Qinghai-Tibet Scientific Expedition Team, as well as part of workers from our Division.

of variegated clastic rocks and limestones, and containing *Burmirhynchia* fauna. The most widespread rocks are the Upper Jurassic to Lowermost Cretaceous flysch and flysch-like formations, intercalated with andesites, andesitic tuffs and minor limestones in the middle part of it, with a total thickness of more than 10,000 m. Eastwards, up to the south of the Lhorong area, they grade into coal measures known as the Duoni coal measures, and westwards, limestones increase. They are unconformably covered by the Uppermost Lower Cretaceous strata containing *Orbitolina* to the west of Nagqu and to the south of Baingoin. The strata below the unconformity have undergone intense compression, producing southerly-overturned tight linear folds whose deformation becomes stronger and stronger towards the Bangong Lake-Nujiang River Fracture Zone. Therefore, the Nagqu Fold Belt is not an Indosinian fold belt but an Yanshanian one.

The northern part of the Nagqu Fold Belt, distributed along the Bangong Lake-Nujiang River Deep Fracture, is a eugeosynclinal belt, making up the second ultrabasic zone. The western segment of this eugeosynclinal belt has been preliminarily ascertained by the Qinghai-Tibet Scientific Expedition Team recently. Pan Yusheng (1981) reported that there is an over-20-km-wide flysch and ultrabasic belt along the Rotog-Gêrzê line, which is undoubtedly the westward extension of the established ultrabasic belt in the Amdo-Baingoin area in the eastern segment.

The southern part of the Nagqu Fold Belt is marked by a granite zone, tentatively called the Nganglha Ring Lake-Lhari Granite Belt. Lying close to the northern side of the Car-Lhari Deep Fracture, it extends from Bowo westwards at least to the Nganglha Ring Lake area, while further westwards it is covered by volcanic rocks. The isotopic ages have three peak values (80—100 m y., 45—60 m y and 10—23 m y). It is evident that this granite zone is a polycyclic magmatic zone of the Yanshanian and Himalayan phases.

The southeastward extension of the Nagqu Fold Belt is named the Tengchong Fold Belt. It is, for the most part, occupied by Gaoligong high-grade metamorphic rocks and Meso-Cenozoic granites and Quaternary volcanic rocks. But considering the fact that the Duoni Formation in t h e Lhorong-Baxoi area in the upper reaches of the Nujiang River definitely extends from Tibet to western Yunnan, we believe that the Gaoligong metamorphic series in the Gongshan area should include not only the Paleozoic but also the Mesozoic rocks. This it is a Yanshanian fold belt as well, while the famous Gaoligong metamorphic zone is most likely to represent the extension of the Gar-Lhari granite zone.

2. *Lhasa Geosynclinal Fold Belt*

As far as the existing data show, the Mesozoic sequence is mainly distributed in the central and western parts of this fold belt, the Lower Triassic having not been found so far. The Middle and Upper Triassic are mainly exposed near Lhasa: the Middle Triassic consists of limestone with andesite and andesitic tuffs, and the Upper Triassic, of sandy shale and limestone, with a total thickness of more than 3000 m, and containing abundant ammonite fossils. Fossil hexacorals, brachiopods and pelecypods have been successively discovered from previously proposed Permo-Carboniferous crystalline limestone, phyllite and schist, which indicates that a certain part of these rocks should belong to the Mesozoic in age. In Mozhugongka, the Jurassic to Lower Cretaceous are represented by coal-bearing clastic rocks, *Orbitolina*-bearing limestone and andesitic volcanic rocks with a considerable thickness. The Upper Cretaceous is separated into upper and lower parts by a regional unconformity: the lower part consists of alternating beds of variegated sandstones and shales, forming tight overturned folds, and the upper part, including the Lower Tertiary, is composed of intermediate-acid (in the eastern part) and intermediate-basic (in the western part) volcanic rocks and pyroclastic rocks with red clastics of continental origin. The tectonic movement that took place in the late part of the Late Cretaceous also represents a main folding episode of the Lhasa Fold Belt.

The famous Gangdisê granite zone lying on the southern margin of this fold belt, is also a Yanshanian and Himalayan polycyclic magmatic zone, where biotite-granite, hornblende granite and grandiorite predominate, their ages being 75—90 my, 40—60 my, and 10—23 my respectively. In the southern foothills of the Gangdise Range, molasses containing Eocene fossils are observed to overlie unconformably the Late Cretaceous-Middle Eocene intermediate-basic volcanic rocks, Yanshanian granites and Himalayan two-mica granite (with an age of 40—45 my).

III.10 THE HIMALAYAN GEOSYNCLINAL FOLD REGION

This fold region is a component part of the gigantic Meso-Cenozoic geosynclinal belt of the Southern Tethys, including the Sulaiman, Himalayan and Arakan Fold Systems. Here, we shall discuss only the Himalayan Fold System, the main part of which is located within the territory of China.

The Himalayan Fold System forms the southern margin of the magnificent Qinghai-Tibet Plateau, adjoining the Gangdisê-Nyainqêntanglha Fold System along the Yarlung-Zangbo-River Deep Fracture on the north. Its main part occurs as an arc slightly convex to the south. Eastwards, it stretches through the south of Zayü and across the border line and then turns abruptly and connects with the Arakan Fold System; while westwards, it runs across the border line and then turns to the southwest in southern Gaparbat and is connected with the Sulaiman Fold System.

The basement rock series making up the Himalayan Fold System is distributed in the main branch of the Himalayas and on their southern slope. Some people regard it as the basement of the Indian Platform. We, on the contrary, consider that it is not the true basement of the Indian Platform but a sedimentary metamorphic series in the Late Proterozoic-Earliest Paleozoic geosyncline on the northern margin of the Indian Platform, with its latest components including not only the Sinian but even part of the Cambrian in its eastern segment. It has become a part of the Indian Platform only since the Middle Cambrian (e.g. the Spiti area in the western segment) and the Ordovician (the eastern segment) (Yin Jixiang et al. 1978).

During the Ordovician to Middle Triassic, the Himalayas were in a state of relatively stable platform. Since the Late Triassic, the tectonic settings have experienced strong differentiation: the southern belt of the Higher Himalayas remained in a relatively stable state, while in the northern belt the activity strengthened and the sediments thickened, taking on the features of flysch formations. In the Northern Himalayas, the Late Triassic, Late Jurassic-Early Cretaceous and Late Cretaceous epochs all witnessed the formation of an ophiolite suite composed of spilite, radiolarian siliceous rocks and ultrabasic rocks. Further northwards, flysch and ophiolite (the Xigazê Group and the Indian flysch) of Late Cretaceous-Eocene ages are distributed along the Yarlung Zangbo River. So the so-called Himalayan Geosyncline is in fact a Meso-Cenozoic geosynclinal belt that has developed since the Late Triassic.

According to the characteristics of rock formations and tectonic deformation, the Himalayan Fold System may be divided, from north to south, into five structural belts: (1) the Yarlung-Zangbo-River Eugeosynclinal Fold Belt, (2) the Northern Himalayan Eugeosynclinal Fold Belt, (3) the Higher Himalayan Nappe Structural Belt, (4) the Lower Himalayan Nappe Structural Belt, and (5) the Siwalik Piedmont Depression Belt.

1. Yarlung-Zangbo-River Eugeosynclinal Fold Belt

The Yarlung-Zangbo-River eugeosynclinal fold belt is precisely the famous Yarlung-Zangbo-River suture. This belt, about 20—40 km across, is largely distributed along the Yarlung Zangbo River, and stretches westwards down the Indus. Thus it is also called by Gansser (1964) the Indus suture. Its eastward extent was not very clear in the past, therefore there have been divergent views about its eastern extension. On the basis of Landsat photographs, combined with analysis of geological data, we have placed its position with certainty to the south of the Médog-Zayü line on the *Tectonic Map of China*, scale 1:4,000,000.

The rock formations within this fold belt consist mainly of flysch and ophiolite of the Xigazê Group and its equivalents. The flysch of the Xigazê Group is intruded by the Gangdisê granite, which is in turn overlain unconformably by the Late Eocene Gangdisê conglomerate (molasse). Thus it is regraded as a Late Eocene fold belt and the product of the final collision between the Indian continent and the Asian continent along the Yarlung-Zangbo-River Deep Fracture Belt.

2. Northern Himalayan Eugeosynclinal Fold Belt

This fold belt borders on the Yarlung-Zangbo-River Eugeosynclinal Fold Belt to the north, and is separated from the Himalayan Nappe Structural Belt by the Lhozhag-Tingri Deep Fracture to the south. This is a eugeosynclinal belt that began to develop in the Late Triassic. The Upper Triassic,

Upper Jurassic-Lower Cretaceous and Upper Cretaceous flysch and ophiolite suites occupy almost the entire geosynclinal belt. On its northern side are widespread mélanges. Before the Late Jurassic a tectonic movement had occurred in this belt, bringing about slight unconformities between the Upper Jurassic and the underlying Upper Triassic or Lower Jurassic, and the lack of the Middle Jurassic in some places (Chen Guoming et al., unpublished data). The over-all folding of the geosyncline also occurred during the Late Eocene Himalayan Movement.

3. *Higher Himalayan Nappe Structural Belt*

This structural belt adjoins the Northern Himalayan Eugeosynclinal Fold Belt' on the north, and is separated from the Lower Himalayan Nappe Structural Belt by the great Central Himalayan Fracture on the south.

It may be divided into the southern and northern subbelts. The southern subbelt is marked mainly by the metamorphic series of the Proterozoic (probably including part of the Cambrian) and the platform cover of the Middle Cambrian to Middle Eocene, and actually it represents the northern margin of the Indian Platform. The northern subbelt, as in the case of the southern subbelt in the Paleozoic, began to develop into a structural transition belt in the Mesozoic, particularly in the Late Triassic, so that the Mesozoic and Lower Tertiary sequences are much thicker than those in the southern subbelt. The highest horizon of the marine sediments also belongs to the Middle Eocene. The early Himalayan Movement occurring in the Late Eocene brought about folding of the sediments. The tectonic movement occurring in the Middle Miocene is the strongest tectonic movement in this belt, accompanied by extensive intrusion of granitic magmas. The most characteristic, large-scale nappe structures in this belt are mostly the products of this age. The nappe structures are generally developed along the bedding planes of incompetent rocks, and the structural planes of the nappes dip to the north as a rule.

4. *Lower Himalayan Nappe Structural Belt*

This belt is adjacent to the Higher Himalayan Nappe Structural Belt on the north and is separated from the Siwalik Piedmont Depression by the Main Boundary Thrust on the south, where mainly the basement metamorphic series of the Himalayas is exposed. The Lower Paleozoic has been only found in the vicinity of Kathmandu, Nepal. It is a metasedimentary-volcanic rock series, and is tightly folded. Thus some geologists consider that a part of this structural belt has undergone the stage of development of the Caledonian Geosyncline. The Permo-Carboniferous is represented by sediments of the Gondwana type. The main period of tectonic disturbance is represented by the late Himalayan Movement, occurring since the Middle Miocene, particularly the Pliocene-Early Pleistocene, which has produced large-scale nappe structure

5. *Siwalik Piedmont Depression Belt*

This depression belt is for the most part located in India, only a small part of it being exposed in China. Its northern boundary is the famous Main Boundary Thrust, and its southern part is buried beneath the Ganges River plain, in which very thick Siwalik molasses were deposited. It is regarded as the result of strong uplifting of the Himalayas and intense subsidence of the piedmont depression at the southern side in Pliocene to Pleistocene times. On the basis of the characteristics of structural deformation, this depression belt may be further divided into the northern subbelt of nappe-imbricate structure and the southern slightly folded subbelt.

III.11 THE MARGINAL-PACIFIC GEOSYNCLINAL FOLD REGION

This fold region refers to a Paleozoic and Mesozoic geosynclinal fold region of the Marginal-Pacific East Asia. It embraces the Variscan Yanbian Fold System and the Mesozoic Nadanhada and Upper Heilongjiang Fold Belts. The Caledonian South China Fold System may probably belong to this fold region, but at present this cannot be affirmed.

III.11.1 The Upper Heilongjiang Miogeosynclinal Fold Belt

The Upper Heilongjiang Miogeosynclinal Fold Belt is located in the northernmost part of Heilongjiang Province and distributed along the Muhe-Deli area of the province. It is considered as a Yanshanian miogeosyncline with the Xingkaiides serving as its basement (the part in the Soviet Union also includes the basement of the Variscides). The geosynclinal sediments are composed mainly of Middle Jurassic sandy shales, nearly 1000 m in thickness. They are largely of continental origin within the territory of China, but become mainly marine in origin when they pass the Heilongjiang Province of China and enter the Soviet Union. It is unconformably overlain by the Late Jurassic-Early Cretaceous continental volcanic rock series, so it is an early Yanshanian fold belt.

III.11.2 The Nadanhada Eugeosynclinal Fold Belt

The Nadanhada Fold Belt is regarded as a part of the Sikhote Fold System. Its Paleozoic folded basement is made up of the Permo-Carboniferous rocks. The Mesozoic geosynclinal deposition started in the Late Triassic and ended in the late part of the Middle Jurassic. The Upper Triassic is called the Zhenjiang Formation, composed of siliceous rocks and tuffs with basalt, about 1500 m in thickness, containing fossil pelecypods (Wang Xiuzhang 1959). The Middle and Lower Jurassic sequences are represented by siliceous rocks, tuffs and black slate with diabase, containing remains of radiolarians, nearly 10,000 m in thickness. The post-geosynclinal sedimentary cover, the Longzhuagou Group, is formed by paralic sediments, containing abundant ammonite and pelecypod fossils.

Here it should be mentioned that a thematic group of the Shengyang Institute of Geology and Mineral Resources, Chinese Academy of Geological Sciences, considers the geosyncline to be not a Mesozoic geosyncline but a Variscan one, as a great abundance of Permian fusulinid fossils has been recently found in the geosynclinal sediments originally assigned to the Mesozoic. At present it is still difficult to make an exact judgement of this contention. On the *Tectonic Map of China*, scale 1:4,000,000, we still assign the Nadanhada area to a Mesozoic geosynclinal fold belt lying on the Variscan folded basement. This assignment is supported by the following lines of evidence: (1) According to Wang Xiuzhang (1959), there exist Mesozoic geosynclinal sediments with fossil evidence in the Nadanhada Mts.; (2) In the Baoqin area west of Nadanhada, the Mesozoic strata with fossil evidence are very thick and are represented by geosynclinal sediments, and the South Shuangyashan Formation is particularly the case, being composed of marine tuffs, tuffaceous sandstone and shale, with a thickness of 2600 m, and containing pelecypod fossils, correlatable to the Zhenjiang Formation of Nadanhada; and (3) Mesozoic geosynclinal sediments with fossil evidence are widespread in the Soviet Union across the Wusuli River, occurring in Nadanhada and to the north of it, facing the outcrops of the Mesozoic geosynclinal sediments in China across the Wusuli River.

III.11.3 The Yanbian Fold System

Located to the east of the Fushun-Mishan Deep Fracture and to the north of the Sino-Korean Paraplatform, the Yanbian Fold System stretches northeastwards to form the basement of the Sikhote-Alin Fold System, and southeastwards into the territory of Korea.

This is a eugeosynclinal fold system, composed mainly of the Permo-Carboniferous. The Carboniferous sequence consists of limestone, slate, tuffaceous sandstone and acid volcanic rocks, with a thickness of 2500 m. The Lower Permian Miaoling Formation is represented by sandstone, sandy conglomerate, siltstone, slate, limestone and intermediate-acid tuffs, 3500 m thick. It rests unconformably or disconformably on the Carboniferous. Both of them contain fusulinid fossils. The Lower Kedao Formation of the Lower Permian consists mainly of gray and grayish-purple tuffaceous conglomerate and tuffaceous sandstone with siltstone, slate and small amounts of limestone lenses, 1000 m in thickness. It lies disconformably on the Miaoling Formation. The Upper Kedao Formation of the Upper Permian is composed mainly of purplish-red and grayish-green marine schistose acid volcanic tuffs, volcanic breccias and lavas with grayish-black sandstone and slate, 850 m in thickness. It overlies conformably the Lower Kedao Formation. In the Hunchun area in the eastern part of the geosyncline, the volcanic rocks are dominated by intermediate lavas, intercalated with basic lavas, with a maximum thickness of 3700 m.

The Yanbian Fold System was originally assigned to the Jilin-Heilongjiang Fold System of the Central Asian-Mongolian Geosyncline. However as the Permo-Carboniferous fossils therein belong to the Pacific warm-water faunas, different from the cold-water faunas in the Central Asian-Mongolian Geosyncline, but closely related to the Late Paleozoic Marginal-Pacific Geosyncline in Sikhote-Alin and Japan, we ascribe it to the Marginal-Pacific Geosynclinal Region.

III.11.4 The South China Fold System

Lying south of the Yangtze Paraplatform, the South China Fold System embraces a wide region from southeastern Yunnan through Guangxi, Guangdong, southern central Hunan, southern central Jiangxi and Fujian to eastern Zhejiang.

This is a late Caledonian geosynclinal fold system. Geosynclinal formations are mainly composed of the Sinian to Silurian rocks, which are represented by flysch and flysch-like sediments, as well as small amounts of carbonate and volcanic rocks. Up to now we have not discovered typical ophiolites of Early Paleozoic age, so it is regarded as a miogeosynclinal belt. Owing to the fact that a large tract of Mesozoic volcanic rocks and granite is distributed in its eastern part, and Devonian-Middle Triassic sedimentary covers are widespread in the western part, the structure of this geosynclinal system in the Early Paleozoic development stage is not yet clear. Whether the South China Fold System belongs to the Marginal-Pacific Geosynclinal Region in the Paleozoic needs further investigation. In recent years, Guo Lingzhi et al. have put forward their idea about the formation and evolution of the South China trench-island arc-interarc basin and foreland basin. This appears to be a subject of research that deserves attention.

The Late Caledonian Movement in the terminal Silurian caused the South China Geosyncline to be transformed into a platform and to merge into the Yangtze Paraplatform. Then the Devonian-Middle Triassic platform cover was deposited. It is largely similar to that of the Yangtze Paraplatform. The difference lies in that the Devonian-Lower Carboniferous sequences in South China are more widespread and thicker than those in the Yangtze Paraplatform.

Since the Late Triassic, the region has entered the stage of development of the continent-marginal activization belt, and has become a very important part of the Western Pacific continent-marginal activization belt. The Indosinian, Yanshanian and Himalayan Movements are all of great importance.

The Indosinian Movement brought about an over-all folding of the Devonian-Triassic sedimentary cover, accompanied by intrusion of granitic and granodioritic magmas. Afterwards, the Anyuan coal measures of the molasse-like type and their equivalents were deposited in intermontane depressions.

The Yanshanian Subcycle was characterized by large-scale volcanic eruption and intrusion of intermediate-acid magma, forming the famous Southeastern China coastal volcanic rock belt and a gigantic granite batholith. Afterwards, very thick gypsum- and salt-bearing red formations and molasse-like sediments were deposited in a number of small down-faulted basins.

During the Himalayan Cycle, the main part of it continued to rise, while the coastal areas were subjected to fracturing and subsidence, accompanied by extensive eruption of basaltic magma.

This must be explained as follows:

1. *Youjiang Indosinian Geosynclinal Fold Belt*

The nature of the geotectonics in the Youjiang area of western Guangxi has long been one of the important problems that have attracted the attention of Chinese geologists since the founding of New China. Throughout many years of geological surveys and petroleum prospecting conducted by geological workers of Guangxi, Guizhou and Yunnan Provinces, this problem has been in the main solved (unpublished data). A brief account is given as follows:

In the Early Paleozoic, the Youjiang area was in the transitional zone between the South China Caledonian Geosyncline and the Yangtze Paraplatform; in the Late Paleozoic, the area was a part of the South China Platform, but more active than the latter; in the Triassic, it was once more transformed into a geosyncline, in which very thick flysch and volcanic rocks were deposited; in the Late Triassic a fold belt was formed by the Indosinian Movement. It is evident that the Youjiang Geosynclinal Fold Belt has in fact constituted part of the Mesozoic Tethyan Geosynclinal Region and lies in the compounding part of the Marginal-Pacific Tectonic Domain and the Tethys-Himalayan Tectonic Domain.

2. *Dongjiang Yanshanian Miogeosynclinal Fold Belt*

As early as the 1950's Triassic-Jurassic marine sediments had been found in the Huiyang, Zijin and Shantou areas, Guangdong Province. Through several years of repeated work it has been proved that they are precisely a sequence of flysch-like sediments with a maximum thickness of more than 6000 m. It is preliminarily considered that it is a minor Yanshanian Miogeosyncline based on the South China post-Caledonian Paraplatform and is most likely to be a "corner" of the gigantic Marginal-Pacific Mesozoic Geosyncline that stretches into the continental margin of China.

3. *Problems of the Nature of the Geotectonics in the Qinzhou and Shiwan Mts. Areas*

Investigations in recent years have demonstrated that in the Qinzhou area the Silurian graptolitic shale and Devonian graptolite-bearing strata are continuous sediments, and that the Late Devonian and Early Permian are marked by a very thick siliceous rock formation. Angular unconformities between the Silurian and the Devonian that can be seen everywhere in South China are absent here. The most important tectonic movements took place between the Early and Late Permian. Very thick Late Permian molasse overlies unconformably older strata including Early Permian ones. Apparently, a Variscan fold belt (Wu Jiyuan 1979) is represented. This fold belt stretches southwestwards and undoubtedly enters Vietnam.

III.11.5　The Southeastern Coastal Fold System

In 1974 Zhang Wenyou et al. (Tectonic Map Compiling Group, Inst. Geol. Acad. Sin. 1974) for the first time assigned the coastal areas of Zhejiang, Fujian and Guangdong provinces to a Variscan fold belt. Afterwards, Chen Bingwei et al. (1978), on the basis of the on-the-spot investigations along the coasts of those provinces as well as in Hainan Island, ascertained that these coastal areas represent a Variscan geosynclinal belt and they suggested moreover that Hainan Island is also a Variscan geosyncline. Both of them are referred to as the Southeastern China Coastal Variscan Fold System. This is precisely the tectonic assignment made in the 1:10,000,000-scale *Geotectonic Map of China* and the 1:4,000,000-scale *Tectonic Map of China*. However some local geological workers from these provinces hold different views on this.

We are preliminarily of the following opinions:

1.　Abundant lithofacies-paleogeographic data have proved that the coastal areas of Zhejiang, Fujian and Guangdong Provinces were an epicontinental region in Devonian to Early Carboniferous times, which are referred to as Cathaysia. Therefore, this region might not be a geosynclinal zones, at least in the Devonian to Early Carboniferous.

2.　In the Permo-Carboniferous, a Variscan eugeosynclinal belt actually existed in the vicinity of Taiwan Island and the Philippines, and was connected with Japan, forming part of the Marginal-Pacific Geosynclinal System in eastern Asia. The miogeosynclinal zone within this geosynclinal belt might also involve the coastal areas of Zhejiang, Fujian and Guangdong, China.

3.　From the data available, it can be seen that there is not enough evidence for determining Hainan Island as a Variscan fold belt. If this Variscan geosynclinal belt existed, it could be a part of the Changshan Variscan geosyncline on the northern side of the Indo-China massif. Perhaps it was not connected with the Taiwan-Phillipines Variscan Geosyncline directly.

III.12　THE GEOSYNCLINAL FOLD REGION OF THE WESTERN PACIFIC ISLAND ARCS

Only a small part of this fold region lies in China, that is, the Taiwan Fold System.

In the Taiwan Geosynclinal Fold System, the oldest strata with fossil evidence are the Permo-Carboniferous sequences in the Danan'ao metamorphic rocks. Therefore, it can be affirmed that the geosynclinal stage started from that period of time. The Danan'ao metamorphic belt may be divided into the high-pressure and low-temperature Yuli metamorphic belt and the high-temperature and low-pressure Tailuge metamorphic belt. These paried metamorphic belts might represent the Benioff Zone along the continental margin of China in the late Yanshanian Cycle, and their northward extensions might correspond to the Rycke and Sanbagawa metamorphic belts. The isotopic age of the Danan'ao metamorphic rocks has been determined to be 86 my, which may represent the age of this movement.

The Taiwan Geosynclinal Fold Region has been separated into two parts since the Himalayan Cycle, with the Danan'ao metamorphic belt serving as a boundary between them. To the east of it is the eastern Taiwan Eugeosynclinal Fold Belt (i.e. the eugeosynclinal fold belt of the Coastal Range in eastern Taiwan), and to the west of it is the western Taiwan Miogeosynclinal Fold Belt and the Taiwan-Penghu Depression.

The oldest strata of the eastern Taiwan Eugeosynclinal Fold Belt are the Miocene Duluan Formation which is composed of marine clastic rocks and volcanic rocks of geosynclinal type and is conformably overlain by the late Miocene-Pliocene Dagang Formation. The Pliocene-Early Pleistocene Taiwan Movement brought the development of the geosyncline to an end.

The deposition in the western Taiwan Miogeosynclinal Fold Belt was initiated in the Eocene. In the Earliest Miocene, the Su'ao Fold Belt originated on the western side of the Danan'ao metamorphic belt. This movement brought about an overall folding and metamorphism of the Paleogene-Early Miocene strata in Taiwan. This movement may correspond to the Takachiho Movement in the Shimanto area of the outer zone of southwestern Japan, as well as to that in the Phillipines. In Taiwan this movement is called the Puli Movement. An isotopic age of 33 my for the Danan'ao metamorphic belt may represent the age of this movement.

In the Neogene the geosyncline migrated westwards. In the Miocene and Pliocene marine clastic rocks of geosynclinal type were predominant, intercalated with several layers of oil- and coal-bearing rock series. They were folded and uplifted through the Pliocene-Early Pleistocene movement. Meanwhile the Taiwan-Penghu Depression originated in the west. Drilling data from Beigang and Penghu have revealed that the Tertiary sequence is underlain by the Jurassic-Cretaceous rocks. The geosyncline might be connected with the continent in the Mesozoic, but divergent views still exist, and further studies remain to be made.

III.13 THE EPICONTINENTAL BASINS AND MARGINAL SEA BASINS IN EASTERN CHINA[11]

The eastern margin of the continent of China is furnished with a vast sea domain, which mainly includes the Bohai Sea, Huanghai Sea, East China Sea and South China Sea. According to their tectonic positions and the salient

---

11 This section is mainly based on the data of the Bureau of Marine Geology and the Division of Airborne Geophysics and Aerogeology under the Ministry of Geology and Mineral Resources.

features of submarine structures, the Bohai Sea and the Huanghai Sea may be ascribed to epicontinental seas, while the East China and South China Seas belong to marginal ones. As geophysical and drilling data obtained from these sea domains are limited at present, here we can only give a brief account of the characteristics of these areas.

III.13.1 The East China Sea

The East China Sea is a Cenozoic sea basin formed since the Himalayan Cycle. According to geophysical data, it may be inferred that there might exist two different kinds of basement. The basement in the west may be composed of Cretaceous volcanic rocks and granite; that in the east, of Mesozoic or older metamorphic series similar to that in the Okinawa Island Arc.

The East China Sea domain may be divided into the following tectonic units of subordinate order (from west to east).

1. *Jizhou-Dachen Uplift Belt*

Lying in the west of the East China Sea, the Jizhou-Dachen Uplift Belt extends from the coastal areas of Zhejiang and Fujian up to the southeastern part of Korea. On the west, the Changle-Xiamen Deep Fracture separates it from the volcanic belt of Zhejiang, Fujian and Guangdong, and on the east the Jizhou-Dachen Deep Fracture separates it from the western East China Sea Depression Belt. The Zhejiang-Fujian volcanic belt is composed mainly of Late Jurassic-Early Cretaceous volcanic rocks, while the Jizhou-Dachen volcanic belt consists mainly of Cretaceous volcanic rocks. Offshore geophysical surveying has confirmed that this belt has a great magnetic gradinet, and that there are NE-and NEN-trending magnetic anomaly zones. These reflect late Mesozoic volcanic rocks, granites and fractures. The formation of magmatic rocks was affected by the late Yanshanian Movement. They are extensively distributed not only in the marginal sea areas of China, but also in Sikhote-Alin of the U.S.S.R. Far East, southeastern Korea and the inner zone of southwestern Japan.

2. *The East China Sea Basin*

The East China Sea basin may be divided into three tectonic units of subordinate order.

a) The Western East China Sea Depression Zone, which is located to the west of the central East China Sea Uplift-Fold Belt and to the east of the Jizhou-Dachen Uplift Belt. Structurally, this depression zone can be connected with a Neogene oil- and gas-bearing basin in western Taiwan. The middle part of the depression consists mainly of Neogene sediments, with a maximum thickness of over 5000 m. This depression zone may belong to a back-arc extensional basin in the central East China Sea Uplift-Fold Belt. Deposition in the depression zone began in the Miocene after the formation of the central East China Sea Uplift-Fold Belt.

b) The Central East China Sea Uplift-Fold Belt is situated between the eastern and western East China Sea Depression Zones. To the south it passes

Diaoyu Island and may be connected with the Su'ao Belt of Taiwan Island formed by the Paleogene sequences; to the northeast it may be connected with the Shimanto area of the outer zone of southwestern Japan (as the Japan Islands have shifted towards Southeast Pacific, now it appears not to be connected with the Shimanto area). According to data obtained by No. 1 Marine Geological Investigation Party (unpublished data), the central uplift-fold belt is composed of the Paleogene and older strata, above which the 3.6—4.4 km/s velocity layer corresponding to the Miocene is mostly absent. According to these data, in connection with the tectonic movement that occurred in the Early Miocene in Japan and Taiwan Island (Known as the Takachiho Movement in Japan and the Puli Movement in Taiwan), it is suggested that the central uplift-fold belt might be an Early Miocene island arc that was formed in this period of time and has not yet emerged from the sea.

c) The Eastern East China Sea Depression Zone, which is bounded by the Okinawa Trough Fracture on the east and adjoins the central uplift-fold belt on the west. The eastern depression belt is elongated, and the topography becomes dramatically steep. It may be divided into the eastern and western parts. The sedimentary features in the western part are similar to those in the western depression zone. There are Miocene and Pliocene reflectors, and the strata have undergone intense folding and fracturing. The eastern part near the trough is filled with a sequence of very thick sediments. According to the velocity layer reflected by seismic refraction profiles, they may correspond to Pliocene-Quaternary sediments. According to gravity data, the value of the Bouguer gravity anomaly in the trough at the eastern side of the eastern depression zone may reach +100 mgal, with a maximum value of +160 mgal in the trough south of Diaoyu Island. The Earth's crust thins markedly. According to Pavlov's calculation (Pavlov et al. 1975), the depth of the Moho disconformity in the East China Sea trough is only 21 km. This shows that this depression belt is situated in a transitional zone from continental crust to oceanic crust.

In sum, the eastern and western East China Sea depression zones show quite different features. Broadly speaking, the western depression zone shows features similar to those of the miogeosynclinal type; the thickness and characteristics of its Neogene sediments resemble those of Neogene miogeosynclinal sediments in western Taiwan. This sequence of strata in Taiwan Island is commonly oil- and coal-bearing, and accordingly the western depression zone is regarded as a very promisng oil- and gas-beasing basin. On the other hand, from the characteristics of volcanic activity, seismic activity, folding and fracturing, as well as the high heat flow values, it can be seen that this zone shows salient features similar to those of a eugeosyncline. As both zones have not emerged from the sea, they may represent geosynclines that are still developing (not yet completed) at present, whereas the central uplift-fold belt may correspond to the Central Range. To put it simply, the western depression may be interpreted as corresponding to a Neogene back-arc extensional basin in the central uplift-fold belt, while the eastern depression may represent a back-arc basin that was formed in a later stage or is being formed in the Okinawa Uplift-Fold Belt.

III.13.2 The Bohai Sea and the Huanghai Sea

Unlike the East China Sea, the Bohai Sea and Huanghai Sea are located in the extensional part of the Sino-Korean Paraplatform and the Yangtze Paraplatform in the sea domain, belonging to epicontinental seas. The Bohai Sea has its foundation on the Sino-Korean Paraplatform. The Late Cretaceous to Tertiary sediments are characterized by an association of basalt-gypsum-salt formation and red clastic formation. Geophysical data indicate that the central Bohai Sea depression zone has a Moho depth of 29 km, lying in a mantle uplift region.

The Huanghai Sea comprises the northern Huanghai Sea and the southern Huanghai Sea. The basement of the former is the Sino-Korean Paraplatform, while that of the latter belongs to the Yangtze Paraplatform. Little information is available on the northern Huanghai Sea, as little geophysical work has been carried out so far.

The southern Huanghai Sea Depression may be correlated with the Northern Jiangsu Depression. The depression began to come into existence in the Late Cretaceous. According to the data of No. 1 Marine Geological Investigation Party, the depression may be divided from north to south into five tectonic units of subordinate order: the Qianliyan uplift, the northern depression, the central uplift, the southern depression and the Funansha uplift. The Qianliyan uplift is separated from the northern depression by a fracture. The magnetic field in the northern depression is characterized by wide and gentle negative anomaly stripes, and the gravity is marked by paternoster-shaped negative anomalies that alternate in parallel. The central uplift and the northern depression are separated by a deep fracture, and so are the southern depression and the Funansha uplift. Zhang Yongxia et al. (unpublished data from the Division of Airborne Geophysics and Aerogeology under the Ministry of Geology and Mineral Resources) hold that the central uplift used to be a depression belt in the Mesozoic, which was uplifted in the Cenozoic, its crystalline basement being a pre-Sinian crystalline massif. The southern depression extends westwards up to the continent, and is connected with the Dongtai Depression of northern Jiangsu there. The magnetic field is mainly represented by positive anomalies, and magnetic bodies are buried to a depth of as much as 4—5 km. According to seismic data, the sediments deposited since the Tertlary may attain more than 4000 m in thickness. In some places the Paleogene deposits may be absent, whereas the Neogene and Quaternary sequences are well developed.

Drilling data have indicated that the two sea areas have undergone three major tectonic movements since the Himalayan Cycle. In the Bohai Sea and the North China area they are manifested by unconformities below the Shahejie Formation (E_3), between the Guantao Formation (N_1) and the Dongying Formation, and between the Pingyuan Formation (Q) and the

Minghuazhen Formation. The three tectonic movements also manifest themselves in northern Jiangsu and the Southern Huanghai Sea area.

The Bohai and Huanghai Seas might have begun to form at the beginning of the Quaternary, during which the Himalayas in western China had just begun to rise in amplitude. While the western part was under intense compressive stresses, the eastern part of China was only in a state of tensile stresses, and it was under such conditions that the Bohai and Huanghai Seas were formed.

III.13.3 The South China Sea

The South China Sea lies in a region where the Marginal-Pacific Tectonic Domain and the Tethys-Himalayan Tectonic Domain intersect, so it shows much more complex tectonic features, that is, it bears the salient features of both the Marginal-Pacific Tectonic Domain and the Tethys-Himalayan Tectonic Domain.

According to the data of the Division of Airborne Geophysics and Aerogeology and No. 2 Marine Geological Investigat on Party under the Ministry of Geology and Mineral Resources, the South China Sea may be divided from north to south into the Guangdong-Guangxi epicontinental depression region, the South China Sea down-faulted region in central sea basin, the Nansha uplift region, and the Xisha uplift region.

1. *The Guangdong-Guangxi Epicontinental Depression Region*

This refers to a wide region between the coasts of Guangdong and Guangxi Provinces and the South China Sea down-faulted region in the central sea basin. The northern boundary is largely represented by the Shantou-Zhanjiang line. The aeromagnetic response is an intensely changeable negative anomaly zone, representing a deep fracture belt, which extends in a NE-direction to meet the Changle-Xiamen Deep Fracture Belt.

The Guangdong-Guangxi epicontinental depression region may be separated from north to south into the northern depression belt, the central uplift belt, and the southern depression belt.

a) *The northern depression belt* is an epicontinental margin depression belt adjacent to the coasts of Guangdong-Guangxi Provinces and is composed from west to east of the Beibu Gulf Depression, the Zhujiang River Mouth Depression and the Taiwan-Penghu Depression. In the depression belt, Cretaceous-Paleogene sediments are commonly developed. Paleocene plankton forminifera have been discovered in drill holes in the western coast of Taiwan, and the Paleocene and Cretaceous along the coasts of the continent usually exhibit the features of continuous deposition; this implies that the formation of the depressions might begin in the Late Cretaceous. This age is consistent with that of the formation of most basins in eastern China. In the Paleogene the fault depression filling-type sediments predominated, while in the Neogene the widespread blanket-type sediments were predominant. The Tertiary sediments may attain a thickness of 2000—4000 m.

b) *The central uplift belt* is a neanly E-W-trending uplift belt composed of the Hainan uplift, the eastern Hainan extension uplift and the Dongsha uplift. These uplifts are all made up of Paleozoic metamorphic series and Mesozoic granites. The burial depth of the bedrock is about 1 km below sea level, and the overlying strata are Neogene in age.

c) *The southern depression belt* is composed of the Yingge Sea Depression, the southeastern Qiongzhou Depression and the northern Xisha Faulted Trough. The Yingge Sea Depression, distributed in a NW-direction, lies in the seaward extension part of the Red-River Fracture Belt. It is a Cenozoic depression with its magnetic bedrock occurring at a depth of more than 6 km. The southeastern Qiongzhou Depression is a newly emerged extensional basin, the burial depth of its bedrock being 3—6 km. The northern Xisha Trough is a rift-type deep trough, with a water depth of over 2 km, flanked on both the north and south by deep fractures, forming an E-W-trending graben. The sediments in the faulted trough are inferred to be Tertiary marine ones, attaining a thickness of 5000 m. The crust is 18 km thick, showing the salient features of transitional type.

2. *The Down-Faulted Region in the Central Sea Basin of the South China Sea*

This is of oceanic crust nature, having a crust thickness of 6—8.7 km and a seawater depth of more than 3000 m. Its basement is formed by oceanic basalts, on which lie only 500-m-thick youngest sediments. The central sea basin represents an extensional basin which is still developing at present. There are still different views as to whether the extensional direction is N—S or NE, but from the existing geophysical and drilling data, it may be inferred that the South China Sea seems to be a ENE-trending extensional sea basin.

3. *The Nanshaand Xisa Uplift Regions*

The basement of the Nansha Uplift may be formed by Precambrian rocks. According to the drilling datta from the Lile beach, a drill hole 4000 m deep has not penetrated the Paleogene sequences. Seismic and drilling data from the shelf areas northwest of Palawan in the Philippines have indicated that there exists an unconformity between the Miocene and the Pliocene. Above the unconformity, the Pliocene and Pleistocene are continuous sediments, while below it the Miocene sequence is folded, which implies the presence of a tectonic movement between the Miocene and the Pliocene.

The basement of the Xishan Islands belongs to continental crust, and is composed of Precambrian rocks, as evidenced by drilling data. The granite gneiss obtained by drilling on Yongxing Island yields an isotopic age of 627 my. It is covered by 28-m-thick weathering crust, and further upwards by Neogene reef limestone and chalk, 1300 m in thickness. This indicates that this region has long been in an uplift environment. Through analyses of the lithology of the rocks and the foraminifer fossils from a drill hole in

Xiyong, Wang Chongyou and He Xiyan (Wang's pers. comm. 1977) believed that there is a depositional hiatus between the middle Miocene and the Pliocence sequences, as is the case with the Nansha Islands. It follows thus that the middle Miocene movement is of great importance to the South China Sea region.

Chapter IV

The Geosynclines of China and Their Main Characteristics

IV.1 THE SUBDIVISION OF THE GEOSYNCLINES OF CHINA AND THEIR DEVELOPMENT

The geosynclinal fold regions of China developed since the Paleozoic belong to three different tectonic domains, namely, the Pal-Asian Tectonic Domain, the Marginal Pacific (Circum-Pacific) Tectonic Domain, and the Tethys-Himalayan Tectonic Domain. All of them are complicated and polycyclically developed geosynclinal systems.

IV.1.1 The Geosynclinal Regions of the Pal-Asian Tectonic Domain

All the geosynclinal regions belonging to this domain are Paleozoic in age. They include the Sayan-Ergun Fold Region, the Tianshan-Hinggan Fold Region, and the Kunlun-Qinling Fold Region. The first two together are also referred to as the Central Asian-Mongolian Geosynclinal Region.

The Central Asian-Mongolian Geosynclinal Region, situated between the Siberian Platform on the one side and the Tarim—Sino-Korean Platform on the other, is very extensive and underwent long-term development from the Proterozoic up to the end of the Paleozoic.

This geosynclinal region as a whole exhibits a subsymmetrical oceanward migration off the continent in a direction predominantly from north to south, and secondarily from south to north (Fig. 20). The Sayan-Ergun Fold Region adjacent to the Siberian Platform is Xingkaiian (Salair) and Caledonian in age. South of it are Caledonides and Variscides. The fold region immediately to the north of the Sino-Korean Paraplatform is Caledonian, and further north Variscan. The geosynclinal region was finally closed up in the areas around North Tianshan, Eren of Inner Mongolia, and Xar Moron (Hsilamulun) River area within the Variscan Tianshan-Hinggan Fold Region. Longitudinally, the geosyncline migrated westwards during the Late Proterozoic—early part of Early Paleozoic times. Its eastern segment around Jiamusi, Ergun, and North Mongolia belongs to the Xingkaiian

Fig. 20 A sketch map showing tectonic migrations in China and adjacent regions.

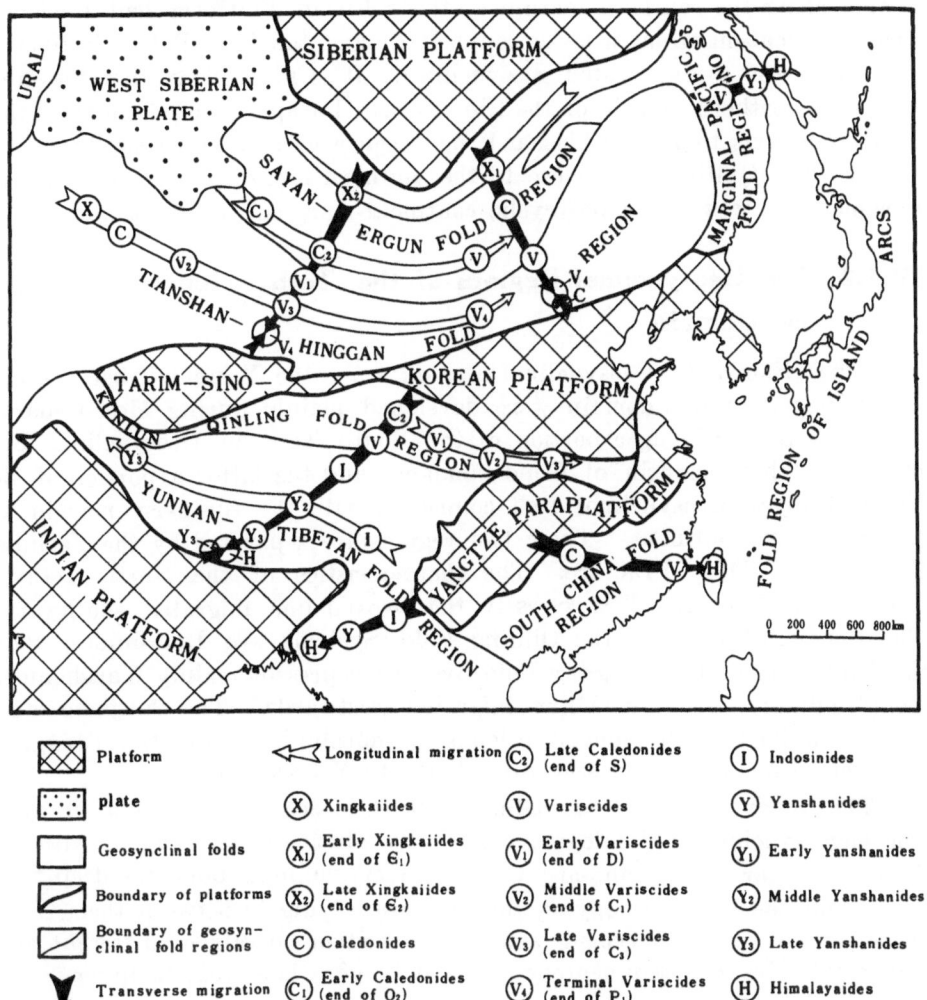

⊠ Platform	⟨ Longitudinal migration	Ⓒ₂ Late Caledonides (end of S)	① Indosinides	
⠿ plate	Ⓧ Xingkaiides	Ⓥ Variscides	Ⓨ Yanshanides	
▱ Geosynclinal folds	Ⓧ₁ Early Xingkaiides (end of Є₁)	Ⓥ₁ Early Variscides (end of D)	Ⓨ₁ Early Yanshanides	
▱ Boundary of platforms	Ⓧ₂ Late Xingkaiides (end of Є₂)	Ⓥ₂ Middle Variscides (end of C₁)	Ⓨ₂ Middle Yanshanides	
▱ Boundary of geosynclinal fold regions	Ⓒ Caledonides	Ⓥ₃ Late Variscides (end of C₃)	Ⓨ₃ Late Yanshanides	
▼ Transverse migration	Ⓒ₁ Early Caledonides (end of O₂)	Ⓥ₄ Terminal Variscides (end of P₁)	Ⓗ Himalayaides	

Fold System, while its western segment at West Sayan and western North
Mongolia is Caledonian. During the late part of the Early Paleozoic—Late
Paleozoic times, the geosyncline migrated eastwards. The western segment
at Junggar and Tianshan started folding early in the Carboniferous, while
the folding of the eastern segment located in Inner Mongolia and Jilin-
Heilongjiang region took place as late as the Permian.

The Kunlun-Qinling Fold Region, situated to the south of the Tarim
Platform and the Sino-Korean Paraplatform, is a geosynclinal region regenerated by taphrogenesis and subsidence of the basement of the Chinese
Protoplatform since the beginning of the Middle Cambrian. It was later
strongly reworked by Tethyan geosynclinal folding, which made the Songpan-Garzê and Qinling regions become actually a part of the Tethyan Geosyncline.

The geosynclines belonging to the Pal-Asian Tectonic Domain are mainly of the Pacific type. Therefore, the early-consolidated Sayan-Ergun Fold
Region along the southern margin of the Siberian Platform had become a
vast continent-marginal activization belt during the Late Paleozoic time (a
part of it was a regenerated geosyncline). It is most likely that the areas
along the northern margin of the Sino-Korean Paraplatform, from the Inner
Mongolian Axis to Alxa (Alashan), had been a coastal range similar to the
Andes. The Qinling Axis close to the southern margin of the Sino-Korean
Paraplatform was also a polycyclic continent-marginal activization belt.

IV.1.2 The Geosynclinal Regions of the Tethys-Himalayan Tecto-
nic Domain

The Himalayan Geosynclinal Region and the Yunnan-Tibet Geosynclinal Region of this domain were developed mainly during Meso-Cenozoic
times, the former being a portion of the South Tethyan Geosynclinal Belt
on the northern margin of Gondwanaland, and the latter a portion of the
North Tethyan Geosynclinal Belt connected with the southern margin of
the Pal-Asian continent. These two regions are separated by the Yarlung-
Zangbo-River Deep Fracture. The characteristics of the North Tethyan
Geosynclinal Belt are: Transversely, the geosyncline migrated southwards.
The Songpan-Garzê and the Qinling Fold Systems are Indosinides, while
the Tanggula and the Gangdise Fold Systems represent mainly Yanshanides.
Longitudinally, the geosyncline migrated westwards. The Sanjiang Fold
System in its eastern segment belongs to Indosinides, while the Tanggula-
Karakorum in its western segment belongs to Yanshanides.

The North Tethyan Geosynclinal Belt is basically of the Pacific type.
The eugeosynclines and miogeosynclines are alternately arranged, with the
former (similar to an island-arc-trench system) running along the deep fractures, and the latter occupying the vast areas lying in between them. Also,
there are several concealed small-scale median massifs of different ages
under the Mesozoic sedimentary cover at Zoigê, Qiangtang, Lanping-Simao,
and Qamdo.

The South Tethyan Geosynclinal Belt is basically of the Atlantic type.
Transversely, the geosyncline migrated away from the continent towards
the ocean in a S—N direction. Its southern part (the North Himalayan
Eugeosynclinal Belt) is mainly a Mesozoic geosynclinal belt, while its northern part (the Yarlung-Zangbo-River Eugeosynclinal Belt) is a Cenozoic geosynclinal belt. Longitudinally, the geosyncline migrated westwards. In the
eastern segment the geosyncline came into being earlier, beginning from
Late Triassic at the latest, whereas in the western segment, geosynclinal
development did not actually take place until the Cretaceous.

In summary, the tectonic migration of the Tethys-Himalayan Geosynclinal Region as a whole also demonstrates a subsymmetrical nature, advancing predominantly from north to south, and secondarily, from south to
north.

IV.1.3 The Geosynclinal Regions of the Marginal Pacific Tectonic Domain

The geosynclinal regions of this domain——the Marginal Pacific Geosynclinal Region, the West Pacific Island-Arc Geosynclinal Region, and the epicontinental and marginal marine basins between them——are mainly Meso-Cenozoic in age. The general trend of development of the geosynclines shows an oceanward migration off the continent in a direction from west to east. The fold systems close to the Asian continent are Paleozoic and Mesozoic in age, and those far from it are Cenozoic. During the Paleozoic, the geosynclines were largely of the Atlantic type. Therefore, no significant influence of the Pacific Ocean upon the Chinese mainland during that time could be observed. However, they were transformed into the Pacific type during the Meso-Cenozoic times, and had gone through a transformation from the Andes type in the Mesozoic to the island-arc—marginal-sea type in the Cenozoic, thus forming an extensive continent-marginal activization belt in eastern China. The West Pacific island-arc belt is a developing or modern geosynclinal region, which furnishes a key to the study of geosynclinal mobile belts, as does the West Atlantic coast of North America.

IV.2 THE CLASSIFICATION OF THE GEOSYNCLINES OF CHINA AND THE SEQUENCE OF THEIR SEDIMENTARY FORMATIONS

Based upon the main characteristics of their tectonic development as stated above, it seems more rational at present to classify geosynclines into the Atlantic type and the Pacific type, and to further subdivide them into eugeosynclines and miogeosynclines.

In the light of the specific features of the geological structures of China, we classify the eugeosynclines into the Qilian type, the Tianshan type, and the Junggar type which combines the characteristics of the former two. The miogeosynclines are further classified into the Bayan Har type, the Qinling type, and the Tanggula type which combines the characteristics of the former two (Fig. 21).

The Qilian-type eugeosynclines, as represented by the North Qilian eugeosynclinal fold belt, are marked by well-developed ophiolitic suites. Based on the studies made by Xiao Xuchang et al. (1978), and on the Stratigraphic Table of Gansu Province, the sequence of formations of the North Qilian eugeosyncline is given as follows:

Devonian	Molasse formation	
Silurian	Arenaceous and argillaceous flysch	>6000m
Upper Ordovician ⎱	Flysch, intercalated with intermediate-basic volcanics	
Middle Ordovician ⎰		>7000m
Lower Ordovician	Intermediate-basic volcanics, represented mainly by basalts (spilites and tholeiites)	2000m
Upper Cambrian	Flysch, mainly sandy and muddy	1200—1800m

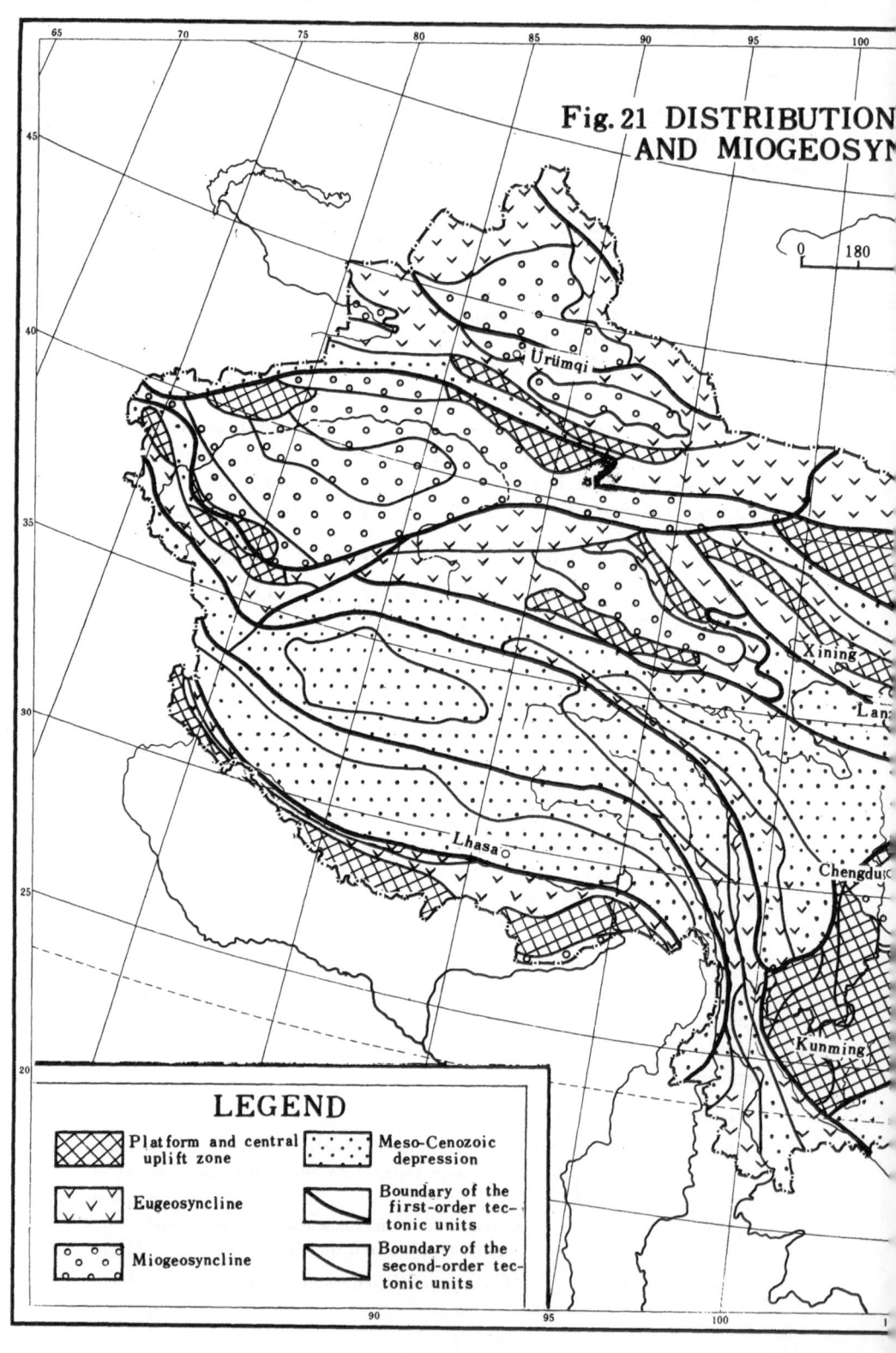

Fig. 21 DISTRIBUTION
AND MIOGEOSYN

LEGEND

⊠	Platform and central uplift zone	⦂⦂⦂	Meso-Cenozoic depression
∨∨	Eugeosyncline		Boundary of the first-order tectonic units
∘∘∘	Miogeosyncline		Boundary of the second-order tectonic units

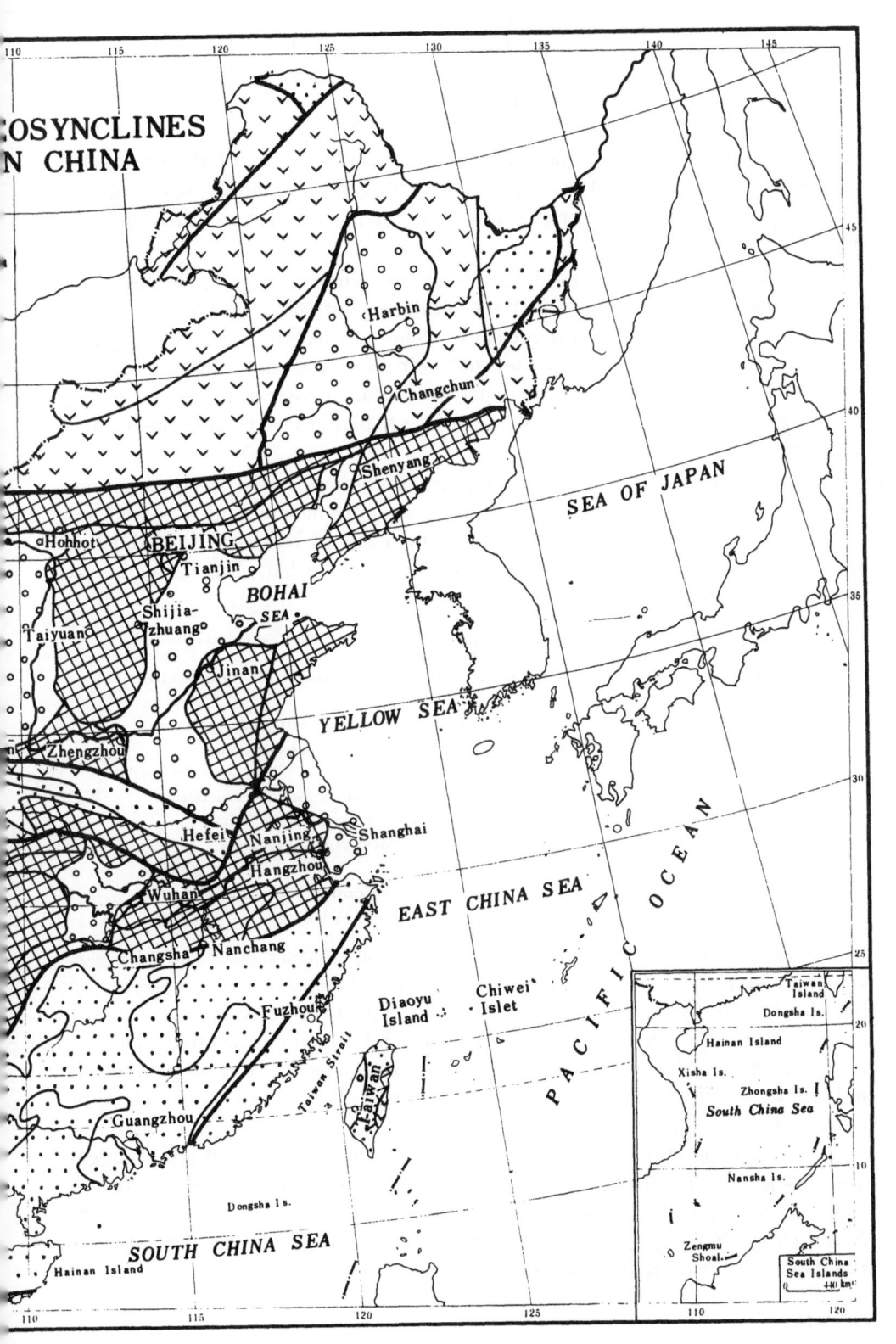

Middle Cambrian
{
Upper part: Volcanics, mainly basic (spilites
and basalts)
Middle part: Predominantly argillaceous flysch,
intercalated with green schist at the bottom
Lower part: gabbro and diabase
(intruded by ultrabasic rocks)
} >3500m

From the above table, one can see that the lower part of the geosynclinal sequence is composed of volcanic rocks, with typical ophiolitic suites, while the upper part consists of terrigenuous clastic deposits overlain unconformably by molasse formation.

The section of the ophiolitic suite in ascending order is: gabbro and diabase→radiolarian cherts and flysch→intermediate-basic and intermediate-acid submarine eruptives→ultrabasics. The ultrabasics were mainly emplaced in a later period, which does not coincide with the ophiolitic section generally reported. This point obviously calls for further studies. In North Qilian Mt., there is also a certain amount of plagioclase-granite associated with the ophiolites.

The Tianshan-type eugeosynclines, as represented by the North Tianshan Eugeosyncline, are characterized predominantly by marine andesites and andesitic clastics. Based on the Stratigraphic Table of Xinjiang (unpublished data), the sequence of formations of the North Tianshan Eugeosyncline is given as follows:

Permian	Molasse formation
Upper Carboniferous	Terrigenous clastics, intercalated with thin-bedded limestones 300m
Middle Carboniferous	Andesite porphyrite, andesite-tuff and tuffaceous sandstone, intercalated with sandstones and limestones 2500m
Lower Carboniferous	Marine intermediate volcanics and volcaniclastics of great thickness, intercalated with basic volcanics, acid volcanics, sandstones and limestones 10000m
Devonian	Marine intermediate volcanics, and volcaniclastics intercalated with carbonates >2500m

From the above table, one can see the main characteristics of the geosynclinal sequence: the lower part is composed of volcanics and volcaniclastics dominated by andesites, and the upper part of terrigenous clastics, overlain unconformably by molasse. It should be pointed out that, judging from the descriptions from other sources, the Tianshan Eugeosyncline not only possesses flysch formation but also several formations of flysch type usually containing volcanic matter.

The Junggar-type eugeosynclines, represented by the West Junggar Eugeosyncline, are characterized by the development of Qilian-type formations in the early stage and Tianshan-type formations in the later stages. Based on the Stratigraphic Table of Xinjiang, the sequence of formations of the West Junggar Eugeosyncline is given as follows:

Middle Carboniferous	Continental volcanics and continental clastics, with conglomerates (molasse) at the bottom	
Upper part of Lower Carboniferous	Neritic-littoral clastics	1400—2200m
Lower part of Lower Carboniferous Upper part of Upper Devonian	Intermediate tuffaceous sandstone and tuff breccia	1000m
Lower part of Upper Devonian	Conglomerate, sandstone, siltstone and thin-bedded limestones	1500m
Upper part of Middle Devonian	Intermediate crystal and detrital tuff, felsite and jasperite	2000—6000m
Lower part of Middle Devonian	(including Lower Devonian) Conglomerate and sandstone	1000—2000m
Upper Silurian Middle Silurian	Tuff, tuffaceous siltstone, andesite porphyrite, basalt-porphyrite, spilite and sandstone	9000m
Lower Silurian	Tuffaceous sandstone and tuffaceous siltstone	1300m
Middle Ordovician Lower Ordovician	Submarine volcanic eruptives, predominantly intermediate and basic (with pillow structure), radiolarian cherts and volcaniclastics	9000m

In the above section, the Lower Ordovician—Upper Silurian are formations of the early stage of geosynclinal development. They are characterized by the development of ophiolitic suites, and flysch deposits. The Middle Devonian (or Lower Devonian)—Lower Carboniferous represent formations of the later stage characterized by marine andesites and andesitic clastics with also the presence of flysch deposits. It should be pointed out that the above is only a section from the Mayile Mt. in western Junggar. In the areas far from it, there are again ophiolitic suites of Late Paleozoic age.

From the three eugeosynclinal sequences given above one can see clearly that they possess a common characteristic: the volcanics and volcaniclastics gradually change, in ascending order, into terrigenous clastics, which are then overlain unconformably by molasse formation.

Based on the classification made by Huang Jiqing et al. (1965), the miogeosynclines of China are further subdivided, according to the amount of clastics and carbonates they contain, into the Bayan Har type, the Qinling type, and the Tanggula type which combines the characteristics of the former two.

The Bayan Har-type miogeosynclines, represented by the Bayan Har Miogeosyncline, are characterized by monotonous lithofacies predominantly flysch deposits in great thickness. Based on the field observations made by Ren Jishun and others in western Sichuan, complemented by the Stratigraphic Table of Qinghai Province, the sequence of formations of this miogeosyncline is given as follows:

Lower Jurassic ⎫	Continental volcanics and coal-bearing	
Upper Triassic ⎬	molasse formation	
Middle-Upper Triassic	Arenaceous and argillaceous flysch formation	>10000m
Lower Triassic	Thin-bedded silty slates and siltstones	80—230m
Upper Permian	Breccia limestone in the lower part and basalt with pillow structures in the upper part	300—400m

The greater part of the Bayan Har Miogeosyncline is made up of Middle-Upper Triassic flysch, best developed in the areas around Jinchuan-Xiaojin in western Sichuan.

The Qinling-type miogeosynclines, represented by the South Qinling Miogeosyncline, are marked by long-term development of carbonate formations. According to the Stratigraphic Table of Gansu Province, the sequence of formations of the South Qinling Miogeosyncline is given as follows:

Middle-Lower Jurassic	Continental coal-bearing formations with conglomerates at the bottom	
Middle-Lower Triassic	Arenaceous and argillaceous flysch formation	>10000m
Permian ⎫ Upper Devonian ⎬	Carbonate formations	2000—5000m
Middle-Lower Devonian	Interbedded carbonates and clastic rocks	3000m
Silurian	Marine volcaniclastics and clastics	

The contact relationship between the Silurian and the overlying strata varies from segment to segment: it may be unconformable in one segment, and disconformable or conformable in another. The amount of volcanic matter in the Silurian sequences themselves also varies from place to place. During the development of the South Qinling Miogeosyncline, the Silurian sequences were already folded, thus forming the marginal uplift of the geosyncline. The Qinling-type miogeosynclines actually commenced with interbedded carbonates and clastics, and, after undergoing a long period of carbonate development, passed on to flysch formations, which were then overlain unconformably by continental coal-bearing formations.

The Tanggula-type miogeosynclines, represented by the Tanggula Miogeosyncline, are characterized by alternating megasequences of carbonates and clastics (sometimes showing flyschoid features) in great thickness. According to the field observations made by Qin Deyu (Qin's pers. comm. 1977), the sequence of formations of the Tanggula miogeosyncline is given as follows:

Cretaceous	Red molasse conglomerates	
Jurassic	Upper Limestone Formation: marl, limestone, intercalated with sandstone	500—1200m
	Upper Sandstone Formation: sandstone, siltstone, intercalated with marl, limestone, and gypsum lenses	600—1100m
	Lower Limestone Formation: limestone, marl,	

	Lower Sandstone Formation:	
	sandstone, intercalated with minor limestone and gypsum lenses	1600—3300m
Upper Triassic	Limestone Formation: limestone, intercalated with clastics	>700m
	Variegated Clastics Formation: purplish-red and greyish-green sandstone, and siltstone	>2000m

The Jurassic sandstones mostly appear purplish-red in colour. Several thousands of meters in thickness, they are often intercalated with flysch formations.

The above six types of China's geosynclines have been classified according to their sequence of formations. Their temporal and spatial characteristics are described as follows:

1. Eugeosynclines of the Tianshan type and miogeosynclines of the Bayan Har type are widespread in China, while the Qilian-type eugeosynclines and the Qinling-type miogeosynclines are relatively less developed. The Tianshan-Hinggan Geosynclinal Fold Region is dominated by Tianshan-type eugeosynclines. Although ophiolitic suites are found in such areas as Junggar, Inner Mongolia and North Tianshan, it is the Tianshan type that dominates in these areas. Except for the Caledonian Qilian Fold System, the Kunlun-Qinling Geosynclinal Fold Region is also dominated by the Tianshan-type eugeosynclines. The Yunnan-Tibet Geosynclinal Fold Region is chiefly occupied by miogeosynclines, with the exception of the areas along the Yarlung Zangbo River, where ophiolitic suites are well developed. Although there are ultrabasic rocks or ophiolitic suites along the Bangong Lake-Nujiang River, Jinsha River-Red River and East Kunlun Deep Fracture Zones, they are generally poorly developed. The South China Geosynclinal Fold Region is dominated by the miogeosynclines of the Bayan Har type.

2. Well-developed ophiolitic suites in China increase in number as field and laboratory studies advance. However, ophiolitic suites of Caledonian age are less in number than those of Variscan age, which in turn are less than those of Alpine age. Caledonian, Variscan and Alpine ophiolitic suites are represented by North Qilian, Inner Mongolia, and the Yarlung Zangbo Geosynclines respectively. The ophiolites in the North Qilian Eugeosyncline are the products of the taphrogenesis in the Chinese Proto-platform during the Cambrian-Ordovician, while those in the Inner Mongolian Eugeosyncline are relicts of the oceanic crust in the Central Asian-Mongolian Geosynclinal Region, their age being Carboniferous-Permian. The Yarlung-Zangbo-River ophiolitic suite, predominantly Cretaceous in age, came into being at the time when marine sedimentation in the Tethys was about to terminate. It should be pointed out that the ultrabasic rocks in northern Qilian Mt. area usually occur in the upper part of the ophiolitic suite section, which is a reversed succession (Xiao Xuchang et al. 1978) quite different from the ophiolitic suite sequence commonly referred to

(Grass 1976).

3. Well-developed ophiolitic suites are generally wanting in the geosynclinal systems within the Chinese Protoplatform, with the exception of the North Qilian Geosyncline. The two major ophiolitic suites of China, one in the Yarlung Zangbo River area, the other close to the border between the People's Republic of China and the Mongolian People's Republic, are both located along the deep fractures on the margins of the Chinese Protoplatform. This implies that most of China's geosynclines are located on the continental or transitional crust, with only a few of them having their basements on oceanic crust.

IV.3   THE FORMATION AND TRANSFORMATION OF THE GEOSYNCLINES OF CHINA

IV.3.1   The Formation of the Geosynclines of China

Viewed from the aspect of their generation, the geosynclines of China developed since the Paleozoic can be classified into two distinct categories: primary geosynclines and regenerated geosynclines. The former refer to geosynclines inherited and developed from those of the Proterozoic, and the latter, to geosynclines regenerated on platform basement since the Paleozoic.

The primary geosynclines——the Central Asian-Mongolian Geosynclinal Region in the north, the ancient Yunnan-Tibet Geosynclin at Region in the southwest, and the South China Geosynclinal Region in the southeast—— are mainly located on the margins of the continental crust of the Chinese Protoplatform.

The Paleozoic Central Asian-Mongolian Geosynclinal Region is a continuation and further development of Proterozoic geosynclines. The Late Proterozoic geosynclines in the areas of Ergun and Jiamusi remained stationary until the beginning of the Cambrian, when folding took place. These folds, together with those in Sayan and North Mongolia, constitute a tremendous Xingkaiian-Caledonian fold system. The Gongyanghe Group and its equivalents in the Baoshan area of western Yunnan within the ancient Yunnan-Tibet Geosynclinal Region are geosynclinal flyschoid formations of Proterozoic-Middle Cambrian age. The geosyncline-type Bei'ao Formation on the northern slope of the Himalayas also belongs to Proterozoic-Middle Cambrian age. This indicates that they are sedimentary formations deposited in Paleozoic geosynclines inherited from the Proterozoic ones. That the Banxi Group and the Sinian-Cambrian flyschoid geosynclinal sediments in the Caledonian South China Geosyncline are continuous in deposition is also a fact commonly recognized by Chinese geologists.

Most of the geosynclines within the continental crust of the Chinese Protoplatform are geosynclines regenerated since the Paleozoic. Generally speaking, there had been closings and regenerations of geosynclines during each tectonic cycle. In the case of China, regenerated geosynclines were chiefly formed during two periods: (1) the post-Xingkaiian, during which the Kunlun-Qinling Geosynclinal Region was formed; and (2) the Indosinian, during which the Yunnan-Tibet Geosynclinal Region and the Youjiang Geosyncline were developed.

Regenerated geosynclines can be formed in two ways: one is gradual transformation, and the other sudden transformation. The former can be exemplified by the western sector of the Tianshan Mts.: in the areas around Tekes south of the Ili Depression and the Sayram Lake to its north, the stromatolite-bearing Upper Proterozoic sequence, which constitutes the basement of the Chinese Protoplatform (Huang Jiqing et al. 1977), is composed of meta-clastics in the lower part and of carbonate rocks in the upper part. Thereupon come the Sinian and Lower Cambrian deposits of the platform cover. The Sinian is composed of tillites, over 120 m in thickness. The Lower Cambrian consists of siliceous phosphorus-bearing rocks, 50 m thick, its characteristic features being similar to those of the Early Cambrian phosphorus-bearing formation in the Yangtze Paraplatform. The Middle-Upper Cambrian consists of alternating siltstones and thin-bedded limestones, 100 m in thickness. The Lower and Middle Ordovician consist of siltstones, siliceous beds, and small amounts of shales and limestones, with a total thickness of more than 1000 m, bearing the marks of transitional deposition. The Upper Ordovician, miogeosynclinal in character, consists of limestones, 1800 m in thickness. The formations ranging from Sinian to Ordovician in the area around Khariktau in southern Tianshan Mts. are in the main similar to those described above, showing also the characteristic features of platform cover, transitional, and miogeosynclinal deposition. The Silurian and Middle—Lower Devonian sequences, however, belong to the Qilian-type eugeosynclinal deposition, and are composed of basic and intermediate lavas, jasperites, intermediate-basic volcaniclastics, common clastics, and limestones, with a total thickness of 10000 m. From here one can see a complete process of gradual transformation of the platform into a miogeosyncline, and then into a eugeosyncline. The sudden transformation type can be exemplified by the North Qilian Geosynclinal Belt, where the Middle Cambrian ophiolitic suites in some areas usually lie unconformably directly upon the folded Proterozoic basement of the Chinese Protoplatform without any transitional beds in between.

As the gradual transformation of the platform into a geosyncline is the result of downwarping, it can also be called downwarped type transformation, during which no remarkable, or only concealed (if any), faulting took place. Sudden transformation can also be termed downfaulted type transformation: it is the result of taphrogenesis, in which faulting is the dominant factor. Although both downwarping and downfaulting originate from deep-seated geological processes, the former generally gives rise to miogeosynclines, as it has no direct connection with the deeper part of the crust; the latter commonly gives rise to eugeosynclines, as large fractures penetrate right into the great depth of the crust, thus forming passages for magmatic activities. Actually, in the process of formation and development of geosynclines, downwarpings and downfaultings are often combined together. Take the already mentioned western segment of the Tianshan Geosyncline for example, the formation of the miogeosyncline in the early stage was obviously dominated by the process of downwarping, while the formation of the eugeosyncline in a later stage was obviously dominated by downfaulting (or taphrogenesis).

IV.3.2 The Transformation of the Geosynclines of China

Like those in other parts of the world, the mode of closure of geosynclines of China is characterized mainly by the sudden transformation, by going through the processes of intense folding, regional metamorphism and magmatic activities. After the cessation of geosynclinal sedimentation, deposition of the platform cover followed. However, as a result of intensified studies, the gradual transformation of geosynclines has now been more and more revealed. That is, the geosyncline rises slowly, with geosynclinal deposits gradually transforming into platform deposits, without going through the process of intensive folding and metamorphism. Such a process is typified in the transformation of the Xingkaiian geosyncline on the northern slope of the Himalayas. According to Yin Jixiang et al. (1978), the Huangdai Formation and the Bei'ao Formation of the Qomolangma Group under the fossiliferous middle and upper parts of the Lower Ordovician on the northern slope of Mt. Qomolangma are continuous in deposition. The Bei'ao Formation is a sequence of geosynclinal flysch deposits which gradate upwards into shallow-water platform deposits. He wrote: "The so-called low grade metamorphic series in the lower part of the Tethys-Himalayan sedimentary sequence comprises chiefly the Sinian to Cambrian including the lower part of the Early Ordovician, which are basically continuous and transitional. Metamorphism in this series decreased in ascending order up to the occurrence of the fossiliferous middle and upper parts of the Lower Ordovician, and the strata above them." "The primary rocks are flyschoid in nature ... deposition during the Sinian period wass generally geosynclinal, with submarine basic volcanic eruption occurring at Kashmir, seemingly characteristic of a eugeosynsline. The sediments had gradually changed into geosynclinal shallow-water deposits during the Late Sinian—Cambrian period, until the early stage of the Early Ordovician, while those during the middle and late stages of the Early Ordovician became, for the greater part, shallow-water platform deposits."

The Tianshan Geosynclinal System in general shows sudden transformation, with the exception of the Bogda area which shows gradual transformation. Here, the Early Permian Lower Jijicao Group is composed of graywackes, tuffs, conglomerates intercalated with siltstone and limestone, and acid volcanics in its lower part, with a thickness of 2000 m; and of graywackes and siltstones intercalated with shale in the upper part, with a thickness of 1700 m. They belong, undoubtedly, to geosynclinal formations. The Upper Jijicao Group of the Upper Permian consists of oil shale-bearing continental clastics, tuffaceous sandstone, tuff, and marl, with a total thickness of 4000 m, representing a transitional formation, from geosynclinal to platformal Lying conformably thereupon are the platform deposits of the Lower Canfanggou Group (upper part of the Upper Permina). From the foregoing, we can also see a complete process of the gradual transformation of a geosyncline into a platform.

In the South China Caledonian Geosynclinal Region proper covering the Hunan, Jiangxi and Guangdong Provinces and the Guangxi Zhuang Autonomous Region, the Early Paleozoic geosynclinal formations experienced strong folding and slight regional metamorphism, and they are overlain with an angular unconformity by Devonian molasse, presenting a sudden transformation. In the marginal parts of the geosyncline, however, where the Caledonian disturbance greatly weakened, a gradual transformation is observed.

Based upon the above analysis, the following rules can be preliminarily summarized in respect to the transformation of geosynclines:

1. At the end of a tectonic cycle or a megacycle, when sedimentation in the most of geosynclines has practically closed, and tectonic movements weakened, the "residual" part of the geosyncline usually shows closure in the manner of gradual transformation, without going through the process of strong folding and metamorphism. Such is the case with the Bogda area in the Tianshan Mts.

2. When there is a lithospheric or translithospheric fracture between the geosyncline and the platform, the transformation of the geosyncline is usually sudden. When there is spatially a tectonic transition between the geosyncline and the platform, the transformation of the former is usually gradual. Both the South China Caledonian Geosynclinal Region (northern margin) and the Xingkaiian geosyncline on the northern margin of the Indian Platform were transformed in this manner.

It seems to be a general rule that the greater the activity within the geosyncline itself, the stronger the tectonic movement at the time of its closure. That is to say, the geosyncline experiences sudden transformation. When the activity of the geosyncline is small and the tectonic movement at its closure is weak, its transformation will be gradual.

IV.4 THE CONTACT RELATIONSHIP BETWEEN THE GEOSYNCLINES AND PLATFORMS OF CHINA

There are two different types of contact relationship between the geosynclines and platforms of China: the abrupt type and the transitional type.

The so-called abrupt contact refers to the connection of the geosyncline with the platform by a lithospheric or translithospheric fracture. For instance, the northern flank of the Sino-Korean Paraplatform is connected with the Inner Mongolian Geosynclinal System by the deep fracture situated on the northern margin of the Inner Mongolian Axis, thus presenting an abrupt contact relationship. Another example is the Yangtze Paraplatform, which is connected in the southwest with the Sanjiang Geosyncline by the Red-River Deep Fracture.

By the transitional type of contact relationship we mean that tectonic transitional zones exist between geosynclines and platforms. Such phenomena can be seen in all the platforms of China. For instance, there is the Jiangnan transitional zone between the southeastern flank of the Yangtze Paraplatform and the South China Fold System, which extends from Zhejiang Province to the district of Youjiang in Guangxi. There are also the Yanyuan-Lijiang and Longmen Mt.-Daba Mt. transitional zones between the northwestern flank of the Yangtze Paraplatform and the Songpan-Garzê Fold System. Lying between the southwestern flank of the Sino-Korean Paraplatform and the Qilian Fold System are the Hexi Corridor transitional zone and the western marginal transitional zone of Ordos. The contact relationship between the Tarim Platform and Tianshan Fold System in the areas around Kelpin is also transitional. The attribution of the tectonic units in these transitional zones depends on the tectonic nature of each unit: those tending to be geosynclinal in nature are assigned to the geosynclinal fold belt. The Hexi Corridor transitional zone, for example, is assigned to the Qilian Fold System on account of its being basically a miogeosynclinal belt; those tending, for the greater part, to be of platform character, are assigned to the platform. For example, the greater part of the Jiangnan transitional zone is assigned to the Yangtze Paraplatform, with only the Youjiang district being assigned to the South China Fold System.

Going deep into the details of analysis, one can see when the geosyncline-platform contact relationship is of the abrupt type, the marginal parts of the platform usually show an Andes-type continental margin, which is typically exemplified by the Inner Mongolian Axis. When the Tianshan-Hinggan Geosyncline was strongly active, the Inner Mongolian Axis was also in great mobility. Hence, not only Caledonian, but also a large amount of Variscan granites are found at the Inner Mongolian Axis. When the platform margin is of the Atlantic type, there appears a tectonic transitional zone between the platform and the geosyncline. Such is the case with the contact relationship between the South China Caledonian Geosyncline and the Yangtze Paraplatform. When the geosyncline-platform contact relationship is something like that between an island arc and a marginal sea, it can also be of the transitional type. Take the Qilian Geosyncline and the Sino-Korean Paraplatform for example: the median uplift of Qilian Mts. is an island-chain uplift, while North Qilian represents a marginal sea-type eugeosyncline. Lying between it and the Sino-Korean Paraplatform is the Hexi Corridor Transitional Zone. Another example is the Tianshan Mts. The median uplift of the Tianshan also represents an island-chain, and the South Tianshan Geosyncline was a marginal sea-type geosyncline during Silurian—Middle Devonian times. Its contact relationship with the Tarim Platform in the areas around Kelpin is also of the transitional type.

To go still further into the problem, the contact relationship between the platforms on the Chinese mainland and their adjacent geosynclines since the Paleozoic can be grouped into two main types: the Pacific type and the Atlantic type. That between the Central Asian-Mongolian Geosynclinal Belt and the Tarim-Sino-Korean Paraplatform belong to the Pacific type, with Alxa and the Inner Mongolian Axis as continent-marginal activization belts. The contact relationship between the Marginal Pacific Geosynclinal Belt and the Chinese mainland belonged to the Atlantic type during Paleozoic times, but in the Meso-Cenozoic, it belonged to the Pacific type. Hence an extensive continent-marginal activization belt was formed in the eastern part of China during this period. The contact relationships between the Kunlun Geosyncline and the Tarim Platform, between the Qinling Geosynclinal Belt and the Sino-Korean Paraplatform, and between the Tethyan Geosyncline and the Yangtze Paraplatform, all belong to the Pacific type. With the development of the geosynclinal belts, continent-marginal activization belts were found in different places at different times. During the Paleozoic, they were located on the southern margin of the Tarim Platform, the southern margin of the Sino-Korean Paraplatform (Qinling Axis and Alxa), and the western margin of the Yangtze Paraplatform (Kam-Yunnan Axis), while during the late Variscan Cycle and Indosinian Subcycle, they were formed at Burhanbuddha, along the Qinling Aixs, the Kam-Yunnan Axis, and also on the southern margin of the Tarim Platform. The Gangdise granitic zone represents an Andestype coastal range running along the southern margin of the then Asian continent in Terminal Cretaceous—Early Tertiary times.

It is precisely on account of the this contact relationship between the continental crust of China and its adjoining geosynclinal belts that the tectonic development of China was chiefly controlled by the development of the gigantic Pal-Asian geosynclinal belts during the Paleozoic, and by the development of the gigantic Marginal Pacific and Tethyan geosynclinal belts during the Meso-Cenozoic.

IV.5   THE POLYCYCLIC DEVELOPMENT OF THE GEOSYNCLINES OF CHINA

Almost all geosynclines of China had gone through a process of polycyclic development. This has been elucidated in many his parpers by Huang Jiqing (1945, 1960, 1978), Huang Jiqing et al. (1965, 1974). Recently, on the basis of summing up relevant data both from home and abroad, Huang has further consummated his theory and proposed a model for the polycyclic development of geosynclinal fold belts (1979). Here a few supplementary remarks on this point are given, based on the investigations made by Jiang Chunfa on the Tianshan-Hinggan and other geosynclinal regions:

IV.5.1   On Tectonic Migration (Fig. 20)

  1. *The Starting Point of Migration*

It is impossible for an extensive geosynclinal belt to be transformed into a fold belt through one single tectonic movement. The transformation of the whole belt can only be accomplished step by step through several movements (each causing only one part of the geosyncline to be folded), which is a process of change from quantity to quality. The segment of the geosyncline that had first been disturbed and folded is the source area of geosynclinal folding, which is designated as the starting point of tectonic migration. From this starting point onwards along the geosynclinal trend, subsequent foldings occur step by step, until the whole geosyncline is turned into a fold belt. This is what we call the longitudinal tectonic migration. Take South Tianshan Mts. for example: folding occurred first in the areas around Kiziltag east of Bosten Lake at the end of the Early Carboniferous, thus bringing the sedimentation to an end in this area, which we call the source point of migration of the South Tianshan Geosynclinal Belt. From this point westwards, in the area of Khariktau west of Bosten Lake, folding took place at the end of the Carboniferous, thus terminating the geosynclinal sedimentation. The sediments therein were then overlain unconformably by thick-bedded Permian conglomerates yielding plant fossils. Further southwestwards along the trend of the geosyncline in the area of Kokshal, the geosyncline was folded at the end of the Early Permian, and the sediments therein were then overlain unconformably by Late Permian continental deposits. The southern Tianshan Mts., up to this time, had been entirely transformed into a fold belt. From the tracks of the three tectonic movements mentioned above, one can see a clear manifestation of longitudinal tectonic migration: folding or uplifting started from one or several points in the geosyncline, and then extended successively along the trend of the geosyncline. Such migrational phenomena can also be seen in the northern Tianshan, Junggar, and Altai geosynclines. It is thus evident that a starting point of migration is common in all geosynclines.

2. *Longitudinal Migration Towards a Definite Direction*

Since geosynclinal folding starts first from one or several points as stated above, here the question arises whether the subsequent movements would take place haphazardly, first in a certain area and then in another. Judging from the data available on the folding of some geosynclines, the answer is negative. As shown by the example of South Tianshan, foldings or uplifts commenced first from the starting point of migration, and proceeded successively along the trend of the geosyncline step by step in a definite direction. That is to say, longitudinal migration of geosynclines proceeds step by step in a definite direction. Such phenomena are true not only with South Tianshan, but also with North Tianshan and other geosynclines (see below). It is a very significant characteristic feature for geosynclinal folding, which should be given adequate attention.

3. *Migration of Eugeosynclines and Miogeosynclines in Opposite Directions*

Eugeosynclines and miogeosynclines, separated by median uplifts are characterized by longitudinal migration in opposite directions. For instance, the South Tianshan Miogeosyncline migrated from east to west (see above), while the North Tianshan Eugeosyncline, on the contrary, migrated from west to east. The western sector of North Tianshan Mts. started folding around the Ili Depression at the end of the Early Carboniferous, thereby bringing the sedimentation in this area to an end. The sediments therein were then overlain unconformably by Middle Carboniferous continental clastics and volcanics. Eastwards up to the area of Bayan Buruk, the Yamansu Formation, composed of volcanics and volcaniclastics, carrying Early Carboniferous marine fossils, several thousands of meters in thickness, is overlain unconformably by continental clastics (the authors suspect there are marine intercalations therein) bearing Middle Carboniferous plant fossils, while this latter possesses a basal conglomerate whose pebbles contain Early Carboniferous fossils. This fully confirms that geosynclinal sedimentation in this area ceased at the end of Early Carboniferous. Further eastwards up to the area of Lesser Burustai, tectonic movement did not bring about a cessation of sedimentation in this area, although there is an unconformity between the Middle Carboniferous and the Lower Carboniferous Yamansu Formation. The Middle Carboniferous sequence (1700 m in thickness) is composed of siliceous beds, andesites, prophyrites, carbonaceous shales, sandstones, and conglomerates whose pebble carry Early Carboniferous fossils. The Middle Carboniferous is overlain unconformably by Late Carboniferous conglomerates. These data indicate that the geosynclinal sedimentation in this area ceased at the end of the Middle Carboniferous. Still further eastwards up to the area of Kumish, the marine volcaniclastics (Dikan'er Formation) of Middle Carboniferous age, 4000 m in thickness, are overlain unconformably by the Aqickbulak Formation with 360 m of purplish red basal conglomerates, containing Permian plant fossils. Further east, in the area of Dikan'er, the Dikan'er Formation and the Late Carboniferous Sumuk Formation (composed of fine sandstone and siltstone, intercalated with thin-bedded limestones, 300 m in thickness) form a continuous sequence. The Sumuk Formation is overlain unconformably by the Early Permian Aqibulak Formation which consists of variegated conglomerates and sandstones, intercalated with siliceous limestones. All this proves that the geosynclinal sedimentation in this area ceased at the end of the Carboniferous. Further eastwards along the trend of the geosyncline towards Beishan, geosynclinal sedimentation ceased at the end of Early Permian. From the above description one can see that folding in the North Tianshan Eugeosyncline started in the west at the end of the Early Carboniferous, with subsequent foldings occurring one after anther at the end of the Middle Carboniferous, at the end of the Late Carboniferous and, finally, at the end of the Early Permian. The direction of its tectonic migration is exactly opposite to that of the South Tianshan Miogeosyncline.

4. *Asymmetry in Transverse Migration*

This is a characteristic feature of transverse tectonic migration which refers to migration of the geosyncline in a direction roughly at right angles to its general trend. When a geosyncline is converted into a fold belt, a new geosyncline would gradually lake shape parallel to the latter. This new geosyncline would again be transformed into a fold belt with another new geosyncline coming into existence again parallel to the latter. Thus the geosyncline continues to migrate transversely. This is commonly referred to in geological literature as the migration of geosyncline. Geosynclines migrating in such a pattern are mostly found and best represented among the marginal geosynclines along the ocean-continent boundary. The geosynclines on the two sides of the ocean continuously give rise to orogenies due to steady compression or subduction of the plates, and the ocean-continent boundary moves step by step in a direction towards the ocean until the final closing of the ocean. The Central Asian-Mongolian Ocean, for example, has its southern margin connected with the Tarim Platform and the Sino-Korean Paraplatform, therefore, the transverse tectonic migration of the geosyncline on this side of the ocean advanced northwards; its northern margin is connected with the Siberian Platform, so the transverse migration on this side is from north to south (see above). Advancing in opposite directions, the geosynclinal migration on the two sides displays a certain degree of symmetry, with the symmetrical axis nearer to the south side. Hence, it is called asymmetrical migration. The extensive geosynclinal belts south of the Tarim-Sino-Korean Platform and north of the Indian Platform also exhibit an asymmetrical migration. Such asymmetry is shown by the area north of the axis: (a) more extensive in areal distribution; (b) having experienced more movements; (c) more Qilian-type eugeosynclines being developed; and (d) more distinct in geosynclinal differentiation. The reasons for these features are no doubt problems worthy of study: they should be probed from such points as the distance between the expanding ocean ridge and the continental margin, the ocean-continent contact relationship and the special features of the activities of the ocean.

5. *Migration off the Uplift Towards the Depression*

The general tened of tectonic migration is that which departs from the uplift twards the depression. It is a very important salient feature not only true for geosynclinal migration, but also true for tectonic migration on the platforms. For instance, at the eastern end of North Tianshan Mts., there is an uplift to the east of Barkol Lake called the Huangcaopo Uplift, which is composed of Early Paleozoic metamorphic rocks. In its vicinity, there are folds of terminal Early Carboniferous age. West of it in the areas of Qijiaojing and Mori (Mulei) County, Late Carboniferous geosynclinal deposits are overlain unconformably by a sequence of continental deposits of Early Permian age. This indicates that geosynclinal sedimentation in these areas ceased at the end of Late Carboniferous. Further westwards in the Bogda area, there rose a landmass at the end of the Early Permian. How did it happen that an east-west migration occurred in this area of North

Tianshan Mts. when the general trend of migration was from west to east? This must be due to the existence of the Huangcaopo Uplift, which brought this part of North Tianshan Mts. under the constraints of migrating away from the uplift to depression. Such a pattern of migration in a direction opposite to the dominant trend of tectonic migration is designated as backward migration. Take the West Junggar as another example: The Mayile and Sarburti fold belts came into being during the Caledonian Cycle. According to field observations by Jiang Chunfa, the intense subsidence of the geosyncline during Middle Devonian—Early Carboniferous times migrated northwards to Sawur Mt. after the folding of the Sarburti Geosyncline, while the latter at this time, commencing with terrestrial conglomerates and sandstones, only showed neritic deposition in the late part of Late Devonian. This northward migration, contrary to the general N-S trend of transverse tectonic migration (in the area north of the asymmetrical axial line), is also a case of backward migration, which was due to the early rise of the fold belt during the Caledonian Cycle.

6. *The Changes of Trend for Different Periods of Time*

Geosynclines migrate in a definite trend during a definite period. Such periods are termed tectonic migration periods. In China, two major periods of geosynclinal migration occurred, namely, the Paleozoic migration and the Meso-Cenozoic migration. When one migration period turns into another migration period, the direction of longitudinal tectonic migration is often transformed into the opposite direction, with the exception of the migration field connecting two periods. This is the nature of change of tectonic migration. The region in which both the longitudinal and transverse migrations trend in their respective dominant and definite directions during a definite tectonic period is designated the tectonic migration field. The region in which two migration periods possess the same migration direction is designated the connective migration field. The region south of the Siberian Platform and north of the Tianshan Median Uplift Belt—Inner Mongolian Axis is called the Central Asian-Mongolian tectonic migration field. It exhibited a west-to-east migration during the Paleozoic, and an east-to-west migration during the Meso-Cenozoic. The latter is reflected by the migration of the center of deposition in the piedmont depression near Urümqi. A similar case is seen in the Variscan Lixian-Zhashui Miogeosynclinal Fold Belt of the Qinling Mts. and in the Indosinian miogeosynclinal fold belt of South Qinling Mts. The former shows a migration from west to east, the latter from east to west. This indicates that when there is a change in the period of tectonic migration, the direction of migration is changed to the opposite.

7. *Residual Character of the Terminal Migration*

As a result of tectonic migration, the geosyncline would have undergone folding step by step until the last stage of its development, when the overwhelming part of it would have been transformed into mountain ranges. The remaining unaffected basins would finally rise slowly to become a land

mass without undergoing the process of folding. Such phenomena we desig-
nate residual tectonic migration. The transitional transformation of the
Bogda Range in eastern Tianshan mentioned above is a typical example.

The above-mentioned seven salient features of tectonic migration, especially the starting point of migration, and the longitudinal and transverse
migrations, are closely related to the polycyclicity of tectonic movement.

IV.5.2 On the Differentiation of Geosynclines

The changes in the regime of tectonic movement and types of formations in the geosyncline are referred to as the differentiation of geosyncline. During its transformation into a fold belt, the eugeosyncline usually undergoes a stage of differentiation, which is characterized by the simul-
taneous occurrence of two different types of formation sequences: the eu-
formation sequence and the mio-formation sequence. In the Junggar
Eugeosyncline, the eu-formation sequence is made up of andesitic volcanics
and volcaniclastics in the lower, volcanic flysch in the middle, and normal
sedimentary rocks in the upper part, while the mio-formation sequence consists of clastics and carbonates. Spatially, these two types of sequences appear as paired zones, as do eugeosynclinal and miogeosynclinal belts. Therefore, the eu-formation sequence, containing volcanic matter, is designated
the eu-formation zone, and the mio-formation sequence, devoid of volcanic
matter, is called the mio-formation zone. They exist together in the eugeosynclinal belts. The concept that eugeosynclinal belts consist exclusively of
volcanics and volcaniclastics does not tally with the facts. The stage of
differentiation can be observed in the Qilian Eugeosyncline after its folding: this corresponds to the stage of development of an island arc. After
the folding of the eugeosyncline, the geosyncline continued to develop, with
a miogeosyncline or mio-formation belt on the continental side, and a eugeosyncline on the oceanic side, with andesites on the island arc. Such a process of differentiation would occur again when there was another tectonic
movement. Each tectonic movement would bring about a differentiation
of the geosyncline. The Qilian-type eugeosyncline was differentiated into a
eu-formation zone and a mio-formation zone, whereas the pre-existing eu-
formation zone can be turned into a mio-formation zone, and the pre-existing
mio-formation zone into a zone of continental deposits. Such differentiation can also be testified by the Junggar Geosynclinal Fold System. From
the above, one can see the evolutionary trend of polycyclic tectonic movement, and it is necessary to take the differentiation of geosynclines into account when dealing with the problem of polycyclicity.

IV.5.3 On the Model of Evolution of Geosynclines —— A Point → Line → Plane Model

As stated above, geosynclinal folding and uplifting start from one or
several points; subsequent folding and uplifting rise one after another in
succession along the trend of the geosynclinal belt in regular order, with

corresponding differentiation, magmatism and mineralization, until the whole geosyncline is transformed into a fold belt which merges with the continent into a single whole. Such is the process of longitudinal tectonic migration, indicating a process of polycyclic tectonic movement. Longitudinal migration is followed by transverse migration, when the ocean-continent contact line moves away from the uplift or contient towards the ocean. New eugeosynclinal and miogeosynclinal zones would then come into being along the new contact line. These newly formed geosynclinal zones, when attacked by subsequent movements, would give rise to new starting points and new longitudinal migration, until the terminal point of migration. A new fold belt would again be formed. Thereafter, transverse migration takes place once again until the entire ocean would finally close up and become an integral part of the land, thus bringing to an end the evolution of the geosyncline. It is apparent that such an evolutionary course of starting point→longitudinal migration→transverse migration is a course of point→line→plane development, just like television scanning which starts first from a point, extending into a line and finally enlarging into an area when the complete image comes into sight. This point→line→plane model, a basic model for geosynclinal evolution, is applicable to most geosynclines. The point→line→plane evolutionary process is also a process of polycyclic evolution, as emphasized by Huang Jiqing.

Chapter V

Deep Fractures and Deep-Seated Structures in China

Based on a comprehensive analysis of geological and geophysical data available, there are shown in the 1:4,000,000-scale *Tectonic Map of China* ten and more translithospheric fracture belts, over eighty lithospheric fracture belts and a still greater number of crustal fracture belts in addition to the regional fracture zones and lineaments reflected in the Landsat photographs, with their tectonic stages ranging from the Proterozoic to the Cenozoic. However, owing to inadequate investigation, we are still unable to explain the whole aspect of the tectonic framework of the Proterozoic and still older deep fractures in China. So here only a brief explanation of the fracture systems since Paleozoic is given (Tables 2, 3, Fig. 22).

V.1 THE TECTONIC FRAMEWORK SHOWN BY DEEP FRACTURES IN CHINA

Deep fractures since the Paleozoic in China can be grouped into three gigantic deep fracture mega-systems. They are the Pal-Asian, the Marginal Pacific, and the Tethys-Himalayan Fracture Mega-systems.

V.1.1 The Pal-Asian Fracture Mega-system

This fracture mega-system comprises a group of deep fractures developed in the Central-Asian-Mongolian Geosynclinal Region south of the Siberian Platform and north of the Tarim-Sino-Korean Platform, and in the Kunlun-Qinling Geosynclinal Region south of the Tarim-Sino-Korean Platform. It is a Proterozoic-Paleozoic deep fracture mega-system which controlled the tectonic development of the Central-Asian-Mongolian and Kunlun-Qinling Geosynclinal Regions as well as the neighboring platforms in Paleozoic time, and can be subdivided into:

1. the Southern Margin Deep Fracture system of the Siberian Platform;
2. the Central Mongolian Deep Fracture system;
3. the Southern Mongolian Deep Fracture system;
4. the Borohoro-Central Tianshan Deep Fracture system;

Table 2. List of deep fractures in China (since Paleozoic)

Mega-system	Name of fracture		Depth	Character	Age of intensive activity	Magmatism and other processes	Remarks
	System	Zone					
Pal-Asian	C. Mongolian	1. Ertix	T	c — s	Pz	Σ, γ	
		2. Derbur	L	c — s	Pz		
	S. Mongolian	3. Almantai	L	c — s	Pz	OS	
		4. Karamaili	L	c — s	Pz	OS	
		5. Eren-Solon	L	c — s	Pz	OS	
	Borohoro-C. Tianshan	6. Northern Margin of Borohoro Mt.	T?L	c — s	Pt(?),Pz, Kz	OS	Belonging to the Tethys-Himalayan Fracture Mega-system in the Cenozoic Era
		7. Northern Margin of C. Tianshan	L	c — s	Pz, Kz	Σ, γ	
		8. Southern Margin of C. Tianshan	L	c — s	Pz, Kz	Σ	
	N. Margin of Sino-Korean Platform	9. Xar Moron	T	c — s	Pt, Pz	OS, gs	
		10. N. Margin of Inner Mongolian Axis	T?L	c — s	Pt, Pz	Σ, γ	
		11. N. Margin of Alxa	T?L	c — s	Pt, Pz		
		12. Shule River	L	c — s	Pz		
	Darbut	13. Darbut	L	s	Pz	OS	
		14. Konggai Alatan	L	s	Pz	Σ	
		15. Sawur	L	s	Pz	Σ	
	Khariktau-Bogda	16. Khariktau-Bogda	L	c — s	Pz, Kz	gs	Belonging to the Tethys-Himalayan Fracture Mega-system in the Cenozoic Era
	Altun-Beishan	17. Qiemo River	L	s	Pz, Kz		
		18. Ruoqiang-Lazhulong	L	s	Pz, Kz	Σ	
		19. Xingxing Xia	L	s	Pz	Σ	
		20. Liuyuan	L	s	Pz	Σ	
		21. Altun	L	s	Pz, Kz	Σ	Belonging to the Tethys-
	Langshan Mt.	22. Langshan Mt.	C	s	Pz, Kz	γ	

continued

Name of fracture			Depth	Character	Age of intensive activity	Magmatism and other processes	Remarks
Mega-system	System	Zone					
Pal-Asian	W. Kunlun	23. N. Margin of W. Kunlun Median Uplift	L	c — s	Pz, Mz, Kz	Σ, γ	Himalayan Fracture Mega-system in the Cenozoic Era
	N. Qilian-N. Huai-yang	24. Longshou Mt.	L	c — s	Pz, Kz	Σ, γ	
		25. N. Margin of N. Qilian	L	c — s	Pz, Mz, Kz	Σ	
		26. N. Qilian	T	c — s	Pt(?), Pz, Mz, Kz	OS, gs	
		27. N. Margin of C. Qilian	L	c — s	Pz, Mz, Kz	Σ	
		28. S. Margin of C. Qilian	L	c — s	Pz, Mz, Kz	OS	
		29. N. Margin of Qinling Axis	T	c — s	Pt, Pz, Mz, Kz	Σ, γ	
		30. S. Margin of Qinling Axis	L	c — s	Pt, Pz, Mz, Kz	Σ, γ	
	N. Margin of Qaidam-N. Huai-yang	31. Danghenan-shan	L	c — s	Pz, Mz, Kz	Σ	
		32. N. Margin of Qaidam	L	c — s	Pz, Mz, Kz	Σ	
		33. Shaliu River	L	c — s	Pz, Mz, Kz	Σ	
		34. Qinghai Lake-N. Huaiyang	T	c — s	Pz, Mz, Kz	Σ, gs, m, γ	
		35. Lintan-Shanyang	T	c — s	Pz, Mz, Kz	Σ, m	
	E. Kunlun-S. Qinling	36. S. Margin of E. Kunlun (Xiugou-Tuosu Lake)	T	c — s	Pz, Mz, Kz	OS, m, γ	
		37. Maqên-Lue-yang	T	c — s	Pz, Mz, Kz	OS, m	
		38. Ankang	C, L?	c — s	Pz, Kz		
	Longmen Mt.-Daba Mt.	39. Longmen Mt.	L	c — s	Pt, Pz, Mz, Kz	Σ	
		40. Chengkou-Fangxian	C	c — s	Pz, Mz		
		41. Niangfan-Guangji	L	c — s	Pz, Mz	Σ	

| Mega-system | Name of fracture | | Depth | Character | Age of intensive activity | Magmatism and other processes | Remarks |
	System	Zone					
Marginal-Pacific	Benioff Zone of W. Pacific	42. Longitudinal Valley of Taiwan	T	s	Mz, Kz	OS, gs, m	
	Southeastern Coast	43. Nan'ao-Hongkong	C	t	Kz	β	Unplotted in map
		44. N. Margin of Zhujiang (Pearl River) Mouth Depression	C	t	Kz	β	
		45. N. Margin of Xisha Trough	L	t	Kz	β, Σ_1	
		46. S. Margin of Xisha Trough	L	t	Kz	β, Σ	
		47. Baoying-Xiangtang	C	c — s	Mz		
		48. Chongming-Jingdezhen	C	c — s	Pz, Mz		
		49. Northeast Jiangxi	L	c — s	Pt, Pz, Mz	Σ	
		50. Chenhuai	C	c — s	Pz, Mz		
		51. Wuchan-Xiaoshan	C	c — s	Mz	γ	
		52. Shaowu	C	c — s	Mz		
		53. Anping-Longquan	C	a	Mz		
		54. Pingxiang-Nancheng	C	c — s	Mz		
		55. Zhongshan-Wenzhou	C	a	Mz	γ	
		56. Lishui-Haifeng	C	c — s	Mz	γ	
		57. Ninghai-Zhangzhou	C	c — s	Mz	γ	
		58. Changle-Xiamen (Amoy)	C	c — s	Mz	γ	
		59. Putian-Lufeng	C	c — s	Mz	γ	

Mega-system	System	Zone	Depth	Character	Age of intensive activity	Magmatism and other processes	Remarks
		Name of fracture					
Marginal-Pacific		60. Qingzhou-Fuqing	C	c — s	Mz	γ	
		61. Zhanjiang-Huilai	C	c — s	Mz	γ	
		62. Jizhou-Dachen	C	c — s	Mz	γ	Unplotted in map
	N. Jiangsu-Huanghai Sea	63. S. Margin of Qianliyan Uplift (Jiashan-Xiangshui)	L	t — s	Pt, Mz, Kz		Only the Jiashan-Xiangshui Deep Fracture is plotted in map
	Tancheng-Lujiang	64. Tancheng-Lujiang	L	s, a	Pt, Mz, Kz	Σ	
		65. Yilan-Yitong	L	a	Mz, Kz	β	
		66. Fushun-Mishan	L	s	Pz, Mz, Kz	β	
		67. Jinxian	L	s	Pt, Pz, Mz	Σ	Belonging to Rimjin-Qingdao Deep Fracture System
		68. Qingdao-Rizhao	C, L?	c—s	Mz	γ	
		69. Guangji-Litang	L	c—s	Mz	Σ	
		70. Susong-Lingshan Mt.	L	c—s	Mz	Σ	
		71. Ganjiang R.	C	c—s	Mz		
		72. Wuchan-Sihui	C	a	Pz, Mz	γ	
		73. Yongxiu-Yulin	C	a	Mz		
		74. Macheng-Jining	C	s	Mz	β	
	Lower Liaohe River-N. China	75. Zhuoxian-Shijiazhuang	C	t — s	Mz, Kz	β	
		76. Xingtai-Anyang	L	t — s	Mz, Kz	β, Σ	
		77. Cangdong	C	t — s	Mz, Kz	β	
		78. Liaocheng-Lankao	C	t — s	Mz-Kz	β	
		79. Changzhi-Nenjiang	L	c — s	Mz		

Name of fracture			Depth	Character	Age of intensive activity	Magmatism and other processes	Remarks
Mega-system	System	Zone					
Marginal-Pacific	Great Hinggan-Taihang-Wuling	80. Yichang-Duyun	L	c — s	Mz	Σ	Their numbers are not plotted in map. Also closely related to the Tethys-Himalayan Fracture Mega-system
		81. Fengdu-Panxian	C	c — s	Mz		
	Fenwei Graben	82. Jiaocheng	C	t — s	Kz		
		83. Taigu	C	t — s	Kz		
		84. Luoyun Mt.	C	t — s	Kz	β	
		85. Huoshan	C	t — s	Kz		
		86. Zhongtiao Mt.	C	t — s	Kz		
		87. Huashan Mt.	C	t — s	Kz		
	NW-trending system in S. China	88. Lianshan-Hongkong	C	c — s	Mz	γ	
		89. Ji'an-dongshan	C	c — s	Mz	γ	
		90. Baokang-Yiyang	C	t — s	Mz		
		91. Nanzhan-Jingmen	C	t — s	Mz		
		92. Zhongxiang-Zhanggang	C	t — s	Mz		
Tethys-Himalayan	Yarlung-Zangbo-River	93. Yarlung-Zangbo R.	T	c — s	Mz,Kz	OS,gs,m,γ	
		94. Lhozhang-Cona	L	c — s	Mz,Kz	Σ	
	Himalayan Overthrust	95. Main Boundary Thrust	C	c — s	Kz	γ	
	Nujiang-Lancang Rivers	96. Bangong L.-Nujiang	T	c — s	Pz, Mz,Kz	Σ,m	
		97. Qieli L.	L	c — s	Mz,Kz	Σ	
		98. Gar-Coqen	L	c — s	Mz,Kz	Σ	
		99. Jiali-Dulong R.	L	c — s	Mz,Kz		
		100. Lancang River	L	c — s	Pz, Mz,Kz	γ,Σ,gs	
		101. Changning	L	c — s	Pz, Mz,Kz	Σ	

Mega-system	Name of Fracture System	Name of Fracture Zone	Depth	Character	Age of intensive activity	Magmatism and other processes	Remarks
Tethys-Himalayan	Jinsha River-Red River	102. Jingsha R.-Red R.	T	c — s	Pt,Pz, Mz,Kz	Σ,OS, gs,m,γ	
		103. Lixian River	L	c — s	Mz,Kz	Σ	
		104. Ailao Mt.-Tengtiao River	L	c — s	Mz,Kz	OS,gs,m,γ	
		105. Gongjue-Qiaohou	L	c — s	Mz,Kz		
		106. Daocheng	L	c — s	Pz,Mz,Kz	Σ	
		107. Garze-Litang	L	c — s	Pz,Mz,Kz	Σ,γ	
		108. Chuxiong-Mengzi	C	c — s	Mz,Kz		
	Daofu-Mashan	109. Daofu-Kangding	L	c — s	Pz,Mz,Kz	Σ	
		110. Yadu-Mashan Mt.	L	c — s	Pz,Mz	Σ	
		111. Baise	C	c — s	Pz,Mz,Kz	β	
		112. Funing	C	c — s	Pz,Mz	β,Σ?	
		113. Wenshan Mt.	C	c — s	Pz,Mz	β	
	Western Shear Fracture System	114. Shiquan R.	L	s	Kz	Σ	
		115. Kashi (Kaxgar) Bangong L.	L	s	Mz,Kz		
	Eastern Shear Fracture System	116. Gyaring L.	L	s	Mz,Kz	Σ	
		117. Damxung	L	s	Mz,Kz	Σ	
		118. Medog	C	s	Kz		
		119. Longmen Mt.	L	c — s	Pt,Pz, Mz,Kz	Σ	Belonging to Pal-Asian Fracture Mega-system in the Paleozoic Era
		120. Lijiang-Anshunchang	L	s	Mz,Kz		
		121. Mongkang-Rangtang	C	s	Kz		
		122. Nanding R.	L	s	Mz,Kz		
	N-S-trending Tension Fracture System	123. Wenquan-Amdo	C	t	Kz		
		124. Tangra Yumco	C	t	Kz		
		125. Dawang	C	t	Kz		

Table 3. List of deep fractures in China (pre-Paleozoic)

Mega-system	System	Zone	Depth	Character	Age of intensive activity	Magmatism and other processes	Remarks
Sino-Korean Paraplatform	Inner Mongolian Axis (N. Margin of Sino-Korean Platform)	126. N. Margin of Inner Mongolian Axis	T	c — s	Pt	Σ, γ	Polycyclic activities since the Proterozoic
		127. N. Margin of Alxa	L	c — s	Pt		
		128. Fengning-Longhua	L	c — s	Pt	Σ	
		129. Jining-Chengde	L	c — s	Pt	Σ	
	Qinling Axis	130. N. Margin of Qinling Axis	T	c — s	Pt	Σ, γ	
		131. S. Margin of Qinling Axis	L	c — s	Pt	Σ, γ	
	Shenmu-Yangyuan	132. Shenmu-Yangyuan					Fractures in the basement of platform revealed by the aeromagnetic surveys
Yangtze Paraplatform	Kam-Yun-nan Axis	133. Lüzhi River	L	c — s	Pt	Σ, γ	
		134. Chenghai	L	c — s	Pt	Σ	
		135. Anning R.	L	c — s	Pt	Σ, γ	Polycyclic activities since the Proterozoic
		136. Xiaojiang R.	L	c — s	Pt		
	Red River	137. Red River	T	c — s	Pt		
	Longmen Mt.	138. Longmen Mt.	T?L	c — s	Pt	Σ, γ	
	S. Margin of Yangtze Platform	139. Jiangshan-Shaoxing	T	c —. s	Pt	Σ	
		140. Xupu-Sibao	L	c — s	Pt	Σ	
		141. Sanjiang (Three-Rivers)	L	c — s	Pt	Σ	
		142. Nanpan R.	?	c — s	Pt?		
Jiamu-si Massif	Mudan River	143. Mudan R.	T	c — s	Pt	Σ, gs, γ	

Explanations to Tables 2 and 3

1. There are 84 lithospheric and translithospheric fractures and 50 crustal fractures in total;
2. The asterisk ✳ denotes a duplicate numbering;
3. T = Translithospheric fractures;
   L = Lithospheric fractures;
   C = Crustal fractures;

c—s = Compression and compression-shear;

t—s = Tension and tension-shear;

s = shear;

a = Alternating compression-shear and tension-shear;

Pt = Proterozoic;

Pz = Paleozoic;

Mz = Mesozoic;

Kz = Cenozoic;

Σ = Ultrabasic rocks;

Os = Ophiolitic suites;

gs = High-pressure metamorphic belt as represented by glaucophane-schists and others;

m = Mélanges

5. the Northern Margin Deep Fracture System of the Sino-Korean Paraplatform;

6. the Darbut Deep Fracture System;

7. the Khariktau-Bogda Deep Fracture System;

8. the Altun-Beishan Deep Frture System;

9. the Northern Qilian-Northern Huaiyang Deep Fracture System;

10. the Northern Margin of Qaidam-Northern Huaiyang Deep Fracture System;

11. the Eastern Kunlun-Southern Qinling Deep Fracture System;

12. the Longmen Mt.-Daba Mt. Deep Fracture System; and

13. the Western Kunlun Deep Fracture System.

The deep fractures in the Central Asian-Mongolian Geosynclinal Region represent themselveslas a series of gigantic arcuate deep fracture systems with the Siberian Platform as a centre of their development. The most important fracture systems and their characteristics are as follows: (1) The Southern Margin Deep Fracture system of the Siberian Platform, located within the territory of the USSR, comprises a series of important deep fractures with the ages of their intensive activities ranging from Proterozoic to Early Paleozoic, and controlled the origin and development of the Proterozoic-Early Paleozoic geosynclinal fold systems (the Beikaliides and Salayirides, i.e. the Xingkaiides) in Baikalia and eastern and western Sayanling on the southern margin of the Siberian Platform; (2) The Central Mongolian Deep Fracture System, whose major part is located within the territory of the People's Republic of Mongolia and extends eastwards into the Greater Hinggan Mts. within the territory of China, where it is known as the Derbur Deep Fracture Belt, while westwards it extends to the Altai Mts. to join with the Ertix Deep Fracture Belt. This is one of the most important deep fracture system in the Central-Asian-Mongolian Geosynclinal Region formed in Paleozoic time and represents a paleotectonic and paleogeographic boundary of great significance; and (3) The Southern Mongolian Deep Fracture System——a southernmost arcuate fracture system that led to the final closure of the Central-Asian-Mongolian Geosynclinal Region, or in other words, it represents the suture line of final collision of the Siberian

Additional material from *Geotectonic Evolution of China,*
ISBN 978-3-642-64874-8, is available at http://extras.springer.com

continental crust with the Tarim-Sino-Korean continental crust. This deep fracture system extends eastwards into the area of Eren and Solon of China, and is known as the Eren-Solon Deep Fracture Zone. Westwards it extends into the area of Xinjiang of China and is known as the Almantai and Karamaili Deep Fracture Zones.

The Northern Marginal Deep Fracture System of the Sino-Korean Paraplatform and the Borohoro-Central Tianshan Deep Fracture System extend in an approximately latitudinal trend. The former includes mainly the Xar Moron Deep Fracture Belt and the deep fracture zones on the northern margin of the Inner Mongolian Axis and on the northern margin of Alxa, while the latter comprises essentially the deep fracture zones on the northern margin of the Borohoro Mt., and on both the northern and southern margins of the Central Tianshan Mts. These two fracture systems are of great importance to the tectonic development of the northern and northwestern parts of China, owing to the fact that the major parts of both are located within the territory of China. They have controlled the development of the Caledonian fold belts of the Borohoro Mt., and north of the Inner Mongolian Axis, and have exerted a great influence on both the Paleozoic paleogeography and the paleotectonic aspect of the Sino-Korean Paraplatform. They have long been an important tectonic boundary in the northern part of China.

It is worthy of mention that the arcuate and E-W-trending fracture systems in the Central Asian-Mongolian Geosynclial Region are intersected, in their western segments, by a series of NE- and ENE-trending fracture systems, among which the most important ones are the Darabut, Khariktau-Bogda, and Altun-Beishan Deep Fracture Systems. In combination with the NW- and WNW-trending fracture systems, they constitute structures of rhombic pattern. These fractures can be clearly seen from small-scale Landsat photographs and from the geological and tectonic maps. They may probably represent a network of fractures formed during the dismembering of the Chinese Protoplatform (Plate XI-1).

Among the deep fracture systems in the Kunlun-Qinling Geosynclinal Region to the south of the Tarim-Sino-Korean Paraplatform, the N. Qilian-N. Qinling-N. Huaiyang, the N. Qaidam-Qinghainanshan-N. Qinling-N. Huaiyang, and the E. Kunlun-S. Qinling Fracture Systems are the three principal fracture systems which controlled the origin and development of the Kunlun-Qinling Geosynclinal Fold Region. The N. Qilian-N. Qinling-N. Huaiyang Fracture System had been active mainly during the Yangtze and Caledonian Cycles and controlled the development of the Yangtze and Caledonian fold belts in the N. Qilian-N. Qinling region. This fracture system extending from the northern margin of the Qaidam Basin through the Qinghainanshan and N. Qinling up to N. Huaiyang, was active mainly in late Caledonian and early Variscan, was revived during the Indosinian subcycle, and controlled the development of the Variscan fold belts occurring along S. Qilian, and northern margin of the Qaidam Basin and N. Qinling (Lixian County-Zhashui County), as well as of the Indosinian fold belts distributed from the Qinghainanshan to the Qinling. The E. Kunlun-S. Qinling Deep Fracture System had been active largely from the late Variscan to the Indosinian stage, and controlled the development of the fold belts in the A'nyemaqen Euogeosyncline and the Indosinian Miogeosyncline in S. Qinling. The Longmen Mt.-Daba Mt. Deep Fracture System constitutes the southern boundary of the Kunlun-Qinling Geosynclinal Fold Region with the Longmen Mt. Deep Fracture Zone having undergone a process of protracted and polycyclic development. The Western Kunlun Deep Fracture System represents a part of the N. Pamir arcuate structure.

V.1.2 The Marginal Pacific Fracture Mega-system

This mega-system comprises a series of fracture systems occurring within a broad area that extends from the Kunming-Yinchuan line easterly to Taiwan Island (West Pacific island arc). It is a complicated fracture mega-system that had been active intensively during the Mesozoic and Cenozoic Eras. The particular fracture mega-system is superimposed upon the Pal-Asian Fracture Mega-system in North and Northeast China and has controlled the development of the Marginal Pacific Tectonic Domain in the eastern part of China. It comprises:

1. the deep fracture system representing the Benioff Zone of the West Pacific island arc;
2. the South China Sea Tension Fracture System;
3. the Southeast Coast Deep Fracture System;
4. the N. Jiangsu-Huanghai Sea Tension Fracture System;
5. the Rimjin-Qingdao Deep Fracture System;
6. the Tancheng-Lujiang Deep Fracture System;
7. the Lower Liaohe River North China Tension Fracture System;
8. the Greater Hinggan Mt.-Taihang Mts.-Wuling Mt. Deep Fracture System; and
9. the Fenwei Graben-Type Tension Fracture System.

The deep fracture system representing the West Pacific Benioff Zone is a most important fracture system in the Marginal Pacific Fracture System, which has long been a natural boundary between the Pacific oceanic crust and the Asian continental crust, and is known as a suture of the West Pacific plate. It is the intensive interaction between the Pacific Ocean and Asian continent along this fracture zone that reformed the tectonic framework of the eastern part of Asia in Paleozoic time, thus creating the Marginal Pacific Tectonic Domain in Eastern China. So it is suggested that the Marginal Pacific Fracture Mega-system in East China is in fact a complicated fracture mega-system with the West Pacific Benioff Zone as a main fracture system. Though some fractures of this mega-system might be fully or partially revived Paleozoic or even Proterozoic ones, they all have originated and developed as a result of the intensive activity of the West Pacific Benioff Zone since the Indosinian Subcycle.

The strong activities of the West Pacific Benioff Zone started some time around the end of the Paleozoic, which might have occurred at different places and in different geological times, thus forming several paired metamorphic belts of different ages and with varying positions in Japan (Miyashiro 1972).

The West Pacific Benioff Zone finds its expression within the territory of China only on the Taiwan Island, which is located at the joint of the Ryukyu island arc with the Philippine island arc. On Taiwan Island, local geologists have discovered reliable evidence for the presence of plate tectonic sutures, such as ophiolite suites, mélanges, glaucophane-schists etc. (Ho 1975).

The South China Sea (a marginal sea) Tension Fracture System, whose domain has been proved by both the domestic and foreign geological and geophysical data (Unpublished data obtained by the No. 2 Marine Geological Investigation Party and Division of Airborne Geophysics and Aerogeology under the Ministry of Geology and Mineral Resources) to be a marginal sea basin with an oceanic crust. Judging from the geophysical data from the Xisha Trough, the submarine rift-valley system in the particular marginal sea basin extends in an approximately EW- or ENE-trend. It is assumed that the recent tectonics and the topographic aspect in South China must have resulted, to a great extent, from the expansion of the South China Sea (as well as of the Philippine Sea).

The Southeast Coast Fracture System refers to a set of predominantly NE- and NEN-trending fracture zones in southeast China, which are, in general, crustal fracture zones. Some of them have existed ever since the Late Paleozoic, but were strongly active in the Mesozoic Era, especially during the Yanshanian Cycle. It was thse fracture zones and the NW-trending fracture zones intersecting them that controlled the massive eruption of volcanoes and intrusion of granitic magma in the Mesozoic, as well as the origin and development of a series of Cretaceous—Paleogene down-faulted basins.

The Rimjin-Qingdao Fracture System refers to the fracture system that is distributed in the areas stretching from the Rimjin River in Korea through the northern part of the Huanghai Sea to Qingdao and Zhucheng in China. Some people suspect that it might be a boundary fracture between the Sino-Korean and the Yangtze Platforms. They also consider that the Rimjin River Fold Belt and the Qinling Fold System used to belong to one and the same fold system which was later separated by the Tancheng-Lujiang Fracture Zone to result in the present-day aspect of their occurrence. To prove this point it is necessary to corry out study in depth in the Qingdao, Zhucheng and Wulian areas.

The Tancheng-Lujiang Fracture System is the most important fracture system in eastern China, and will be discussed especially in the ensuing paragraphs.

The Lower Liaohe-North China Tension Fracture System and the Northern Jiangsu-Huanghai Sea Tensile Fracture System refer to those tension fracture systems that are located in the areas of northern Jiangsu, Huanghai

Sea, Lower Liaohe River and North China, and are closely associated with the origin and development of the post-Yanshanian down-faulted basins. Among them the Cangdong and Liaocheng-Lankao Fracture Zones can be regarded as typical representatives. Along the fractures are found eruptions of basalts. As is proved by geophysical data, all these fractures have cut through the Moho (Teng Jiwen 1974).

The Greater Hinggan Mts.-Taihang Mts.-Wuling Mt. Fracture System comprises the Changzhi-Nenjiang and Duyun-Yichang Fracture Zones. That part of the Changzhi-Nenjiang (Zijingguan) Fracture Zone which is seated within the Sino-Korean Paraplatform may probably be a fossil fracture belt of Archeozoic age, as proved by the distribution of Archeozoic ultrabasic rocks along the belt. The modern Zijingguan Fracture and its extension to the Greater Hinggan Mts., however, have resulted from the strong activity of the Pacific Tectonic Domain in Mesozoic times. The Duyun-Yichang Deep Fracture Zone starts from the China-Vietnam border line in the south and extends northwards through Duyun, Zhenyuan and Yongshun areas to disappear in western Yichang area, while southwards it enters the territory of Vietnam. The particular fracture belt is demonstrated clearly in the Landsat photographs. The position of the Greater Hinggan-Taihang Mts.-Wuling Mt. Fracture System is basically coincident with that of the Greater Hinggan-Taihang-Wuling gradient belt of gravity anomalies. It is this very fracture system that separates the Marginal Pacific Tectonic Domain in eastern China into the eastern and western regions with their different history of development. Nearly all the down-faulted basins that occurred during the Mesozoic and Cenozoic times in eastern China are distributed to the east of this fracture system, the Fenwei Graben-Type Down-faulted Basin being an exception.

The Fenwei Graben-Type Tension Fracture System refers to a tension fracture system that extends for over 1000 km from Baoji, through Xi'an, Huaxian, Linfen, Taiyuan, Xinxian, Datong and up to Huailai to form the well-known Fenwei Graben, where there are accumulated very thick Cenozoic (essentially Neogene and Quaternary) clastic rocks of continental origin. Along the edges of the graben there is a series of tension fractures with dextral shearing, while the central part of the graben is characterized by upwarping of mantle (Institute of Geophysics, Academia Sinica 1974) with frequent basaltic eruptions. From the thickness of Cenozoic deposits in the graben it can be seen that its southern segment, the Xi'an-Taiyuan area, is characterized by thicker deposits, greater mobility and earlier formation of the deposits, with that part in the Xi'an-Yuncheng area being formed as early as in Paleogene, while its northern segment, the Datong, Yenqing and Huailai areas, is characterized by thinner deposits and smaller mobility. The Fenwei Graben-Type Fracture System represents an important area of earthquake activities in North China. It is sometimes compared with the Baïkal Rift-Valley. However, it seems that more reliable practical data are wanting.

V.1.3 The Tethys-Himalayan Fracture Mega-system

This is a Meso-Cenozoic fracture mega-system in southwestern China with a series of gigantic arcuate fracture systems as the main. Joining with the Pal-Asian Fracture Mega-system in the Kunlun-Qinling Geosynclinal Fold Region, it has controlled the development of the gigantic Tethys-Himalayan Geosynclinal Region, causing some of the fossil fractures in the Pal-Asian continent to revive. This fracture mega-system can be subdivided into:

1. the Yarlug-Zangbo River Deep Fracture System;
2. the Himalayan Overthrust System;
3. the Nujiang R.-Lancang R. Deep Fracture System;
4. the Jinsha River-Red River Deep Fracture System;
5. the Western Qinghai-Tibet Shear Fracture System;
6. the Eastern Qinghai-Tibet Shear Fracture System;
7. the Central Qinghai-Tibet Tension Fracture System;
8. the Yulong-Longmen Deep Fracture System; and
9. the Kam-Yunnan-Axis Deep Fracture System.

The gigantic Yarlung-Zangbo River, Nujiang-Lancang Rivers, and Jinsha River-Red River Arcuate Fracture Systems are main fracture systems of the Tethys-Hima'ayan Fracture Mega-system, which represent respectively the sutures of various stages of the development of the Tethys-Himalayan Geosynclinal Fold Region. Of them, the Jinsha River-Red River Fracture System is essentially of the Indosinian in age; the Nujiang River-Lancang River Fracture System, mainly of Yanshanian age; and the Yarlung-Zangbo River Fracture System, largely of Himalayan age.

The Eastern Kunlun-Southern Qinling Deep Fracture System and the Northern Qaidam Margin-Qinghainanshan-Northern Qinling-Northern Huai-yang Deep Fracture System represent also sutures of the Indosinian geosyn-clinal fold systems. The difference lies only in that they are furnished with a dual state of being a member of the Pal-Asian Fracture Mega-system at the same time representing an element of the Tethys-Himalayan Fracture Mega-system, the latter being the inheritance and development of the former.

It should be pointed out hereby that although, as is the Yarlung-Zangbo River Deep Fracture Belt, the Qinghai Lake-Northern Huaiyang, Eastern Kunlun, Jinsha River-Red River, and Nujiang River-Lancang River Deep Fracture Belts are all translithospheric fracture zones of a sutural nature, they are quite different from the former both in scale and significance. The Yarlung-Zangbo River Deep Fracture Belt is a first-order suture of global nature, a suture between gigantic geoblocks (plates); while the Jinsha River-Red River, Nujiang River-Lancang River, Eastern Kunlun, and Qinghai Lake-Northern Huaiyang Deep Fracture Belts are sutures between small-scale geoblocks on the southern margin of the Eurasian continent with only local significance. The regions where these fractures happen to occur re-present island arc-trench systems of oceanic crustal or of transitional nature.

The areas lying between these deep fracture belts, such as the Qiangtang Massif in northern Tibet, the Roigê Massif in western Sichuan, and Qamdo are regions with continental crust.

Another point that must be stressed is the importance of the Yarlung Zangbo River Deep Fracture System. The final drastic subduction of the Tethyan oceanic crust towards the Eurasian continent took place in the late stage of the Late Cretaceous (80 my ±). The closure of the Tethys Sea and the collision between the Indian subcontinent and the Eurasian continent occurred in the Late Eocene (45 my ±). The Indian subcontinent and the Eurasian continent were then merged into a single whole. Since then, owing to continuous expansion of the Indian Ocean, the Indian subcontinent kept compressing towards the Eurasian continent, thus giving rise to the Himalayan Overthrust System to the south of the Yarlung Zangbo River, and to the NE- and NW-trending shear fracture zones and approximately N-S-trending tension fracture zones in the Qinghai-Tibet Plateau. Some geologists consider the Main Boundary Thrust of the Himalayan Overthrust System as a suture. However, it is in fact not deep enough to be a translithospheric fracture, but only a crustal one.

The shear fracture system in the western part of the Qinghai-Tibet Plateau is dominated by N-W-trending dextral shear fracture belts represented by the Shiquan River Deep Fracture Zone, which refers to a deep fracture belt extending northwestwards along the Shiquan River. Some people have mistaken it for the Yarlung-Zangbo River Deep Fracture Zone in western Tibet on account of its close proximity to the latter. In the eastern Qinghai-Tibet Plateau, NE-trending shear fracture system predominates, which can be divided into: the NE-trending sinistral shear fracture zones that are located to the west of the major bend of the Yarlung Zangbo River and are represented by the Damxung Fracture Belt; and the NE-trending dextral shear fracture zones that are located to the east of the major bend of the Yarlung Zangbo River and extend from the river bend up to the Yulong Mt.-Longmen Mt. area, with the Yulong-Longmen Deep Fracture System as its major southeastern boundary. The difference in the shearing direction expressed by a NE trend of the shear fractures in eastern Qinghai-Tibet Plateau has resulted chiefly from the effects of a projecting terrain at Xilong at the northeastern end of the Indian Platform (Huang Jiqing 1945). The N-S-trending tension fracture system in central Qinghai-Tibet Plateau represented by the Tangra Yumco and the Wenquan-Amdo Fractures which cut across the Himalayan and the Yunnan-Tibet Geosynclinal Fold Regions. These fractures are filled with Cenozoic intermediate and basic volcanic rocks and form huge hot spring zones.

The Kam-Yunnan-Axis Deep Fracture System which comprises the Anning River, the Lüzhi River and the Xiaojiang Deep Fracture Belts is a deep fracture system that has been active ever since the terminal Proterozoic. Being situated on the western margin of the stable region of eastern China, and represented essentially by platforms, it was active from Paleozoic to

Early Mesozoic due to the strong influence of a geosynclinal mobile region to its west. It has become an extremely important tectono-magmatic metallogenetic zone, at least since the Early Variscan Cycle. During the Mesozoic, especially since the Himalayan Cycle, owing to the strong influence of the Tethys-Himalayan Tectonic Domain, neotectonic activities were exceptionally intense and earthquakes were frequent. In the area of Yuanmou and other places in the country, the Triassic—Jurassic sequences can be found overthrusting on the Quaternary strata (Zhao Guoguang and Pu Qingyu's pers. comm. 1965). Again, having reached the area of Fuxian Lake, the Xiaojiang Fracture Beit has disappeared and can be found nowhere on the surface, while the Landsat photographs show that it has reached the area of Gejiu (Plate XII).

V.1.4 An Interim Summary

Based on the descriptions given above, a brief account of the fracture framework since the Paleozoic of China can be summarized as follows:

The Pal-Asian Fracture Mega-system represents a Proterozoic-Paleozoic fracture mega-system which has controlled the evolution of the Pal-Asian geosyncline. It had been of great importance to the development of the geotectonics of China during the Paleozoic times. The Tethys-Himalayan and the Marginal Pacific (Circum-Pacific) Fracture Mega-systems are Mesozoic and Cenozoic in age. They have controlled the development of Meso-Cenozoic tectonics and determined their characteristics. Some fracture systems from the Tethys-Himalayan Fracture Mega-system have been compounded with those from the Pal-Asian Fracture Mega-system within the Kunlun-Qinling Geosynclinal Fold Region to give the fractures in this region a compound nature. The Marginal Pacific Fracture Mega-system has been compounded and intercrossed with the Pal-Asian Fracture Mega-system in North China and Northeast China. But their relationships are characterized by an evident intersection nature owing to the fact that they have a completely different tectonic framework. This kind of compound relationship is quite different from that between the Paleozoic and Meso-Cenozoic fracture systems within the Kunlun-Qinling Fold Region. It is the origin, development, intersection and compounding of these three fracture mega-systems that have controlled the tectonic development of China since the Paleozoic, and have determined the salient features of the geological structures and the modern geomorphology of China.

It was formerly thought that the boundary between the Marginal Pacific (Circum-Pacific) and the Tethys-Himalayan Fracture Mega-systems was located in the Kunming-Yinchuan area (Huang Jiqing et al. 1977). It is indeed an existential fact that the Kunming-Yinchuan tectonic line did represent a boundary between a rigid geoblock composed essentially of platforms in eastern China and a flexible geoblock consising mainly of geosynclines in western China in Paleozoic and Early Mesozoic (Triassic) times, and this assumption has been universally accepted by the Chinese geologists.

However, form the viewpoint of a fracture system, the fracture belts extending along the Kunming-Yinchuan area are in no way a unified fracture system, and the various fractures belong respectively to different systems. Judging from their salient features and history of development, these zones must be considered as belonging to the Tethys-Himalayan Fracture Mega-system. The Marginal Pacific Fracture Mega-system is distributed mostly to the east of the Sichuan and the Ordos (within the Territory of China), tremendously stable rigid massifs. Consequently, it would be more correct to say that the Sichuan and Ordos rigid massifs have separated the Marginal Pacific (Circum-Pacific) and the Tethys-Himalayan Tectonic Domains from one another rather than that the Kunming-Yinchuan line used to serve as a boundary between the eastern and western parts of China in the Mesozoic and Cenozoic eras.

V.2 BRIEF DESCRIPTIONS OF THE MAIN DEEP FRACTURES IN CHINA

V.2.1 Translithospheric Fracture Zone

Within the territory of China, there are ten and more translithospheric fracture zones hitherto determined or preliminarily determined according to the data available. They are of great importance to the study of the geotectonic development, the distribution of mineral resources, and the occurrence of earthquakes in China.

1. *The Gigantic Taiwan Longitudinal-Valley Deep Fracture Zone*

This fracture zone lies between Hualian and Taidong in the eastern part of Taiwan with its exposure extending for 150 km. On its eastern side runs a eugeosynclinal fold belt along the Taiwan Coastal Range, while on its western side is the Danan'ao metamorphic belt. Distributed along the fracture belt are ophiolitic suites, mélanges (the Pliocene-Pleistocene Lichi Formation) and glaucophaneschist (Yuli Belt) (Ho 1975). It represents part of the Benioff Zone of the West Pacific Cenozoic island arc——a suture between the Asian continent and the Pacific Ocean.

2. *The Yarlung-Zangbo River Deep Fracture Zone*

This is a most attractive deep fracture zone in the western part of China, and is considered as being a suture between the Indian and the Eurasian plates. It extends within the territory of China for 2000 km.

This fracture zone runs essentially along the Yarlung Zangbo River and extends westwards along the Indus River into India and Pakistan, which is shown clearly on the geological maps and Landsat photographs, and has long been recognized by Chinese geologists. But its easterly extension, after passing through the area of Langxian County, has been unknown due to intensive magmatism and metamorphism of the rocks at the major bend of the Yarlung Zangbo River, and to the complicated relief and poor natural conditions for geological work. In 1974, Huang Jiqing et al. made a preliminary analysis of it on the basis of the geological data available and came to an understanding that this fracture zone does not enter the territory of Yunnan Province when it reaches the major bend of the Yarlung Zangbo River, but turns southerly to enter Burma and reaches the Uyu Valley after its passing through the Zayu area. Further southwards, it becomes an ophiolite zone in the Arakan Yoma. The analysis of the Landsat photographs has completely proved this suggestion. On the Landsat photographs there are clear signs showing that this fracture belt extends through the Medog County up to the south of the Walung area in Zayu County, and then turns from the major bend of the Zayu River to enter the territory of Burma, where it joins with the ophiolite zone in the Putao (Fort Hertz) area (Plate X). What is most significant is that, as reported by Sinkha (1977), an ophiolitic zone and a high-pressure and low-temperature metamorphic belt were found to the south of Walung and along the northeastern side of a NW-trending valley at the Zayu River bend. North of this high-pressure metamorphic belt lies a high-temperature and low-pressure metamorphic belt composed of the Zayu Granite, which is undoubtedly the easterly extension of the Gangdisê granitic belt. Thus, we have completely confirmed in several respects the exact position of the easterly extension of the Yarlung-Zangbo River Deep Fracture Zone.

Along the Yarlung-Zangbo River Deep Fracture Zone is distributed the largest ophiolite belt known so far within the territory of China. In addition to the well-known mélange at the Qiargar Peak, the Qinghai-Tibet Scientific Expedition Team from Academia Sinica, and Xiao Xuchang and Wang Naiwen from the Institute of Geology of the Chinese Academy of Geological Sciences have found in recent years several occurrences of mélanges in a 700-km-long segment of the particular fracture zone extending from Zhongba up to the area east of Yamzho Yumco. On the northern side of the fracture there lies a 2000-km-long Gangdisê granitic belt, while on its southern side a glaucophane-facies, high-pressure and low-temperature metamorphic belt is found in the areas of the Zayu River valley (the eastern segment) (Sinkha 1977), and the Indus River valley (the western segment) (Jan et al. 1973). The authors are confident that further investigations will certainly result in more discoveries.

3. *The Bangong Lake-Nujiang River Deep Fracture Zone*

The Bangong Lake-Nujiang River Deep Fracture Zone is a gigantic arcuate deep fracture zone which starts from the Bangong Lake in the west, passes through Gerze, Lunpuolha and Sogxian County, turns southwards when it reaches Dengqen County, and then proceeds onwards along the Nujiang River. To the west, it enters India and Pakistan, while to the south, it extends up to Longling County in western Yunnan where it turns to the southwest to enter Burma along the Ruili River. It extends for 2800 km within the territory of China, constituting the dividing boundary between the Indosinian-Early Yanshanian Karakorum-Tanggula-Sanjiang Fold System and the Middle-Late Yanshanian Gangdisê-Nyainqentanglha Fold System. The authors deem it necessary to explain here that the Lancang River Fracture Zone was formerly prolonged up to the Bangong Lake through Dengqen, Sogxian and Amdo Counties; but it now seems that the Nujiang River Fracture Belt is a more important fracture, which meets with the Bangong Lake-Sogxian Deep Fracture Belt in the vicinity of Dengqen County. This proves that the original delineation made by the Geological Bureau of the Tibet Autonomous Region is more consistent with the fact. As for the Lancang River Deep Fracture Zone, one branch of it meets with the Bangong Lake-Sogxian Fracture to the west of Qamdo after it reaches the eastern piedmont of the Taniantaweng Mt., while another branch continues to extend northwesterly up to the south of the Kaixin Ridge through the Danta area.

The Bangong Lake-Nujiang River Deep Fracture Zone has always been an intensively active and polycyclic tectono-magmatic and metamorphic belt since the Indosinian Subcycle. In the western segment, its southern side represents a strongly compressed and folded green schist belt composed of graywacks, volcanics and siliceous rocks, while its northern side is a high-temperature metamorphic belt with andalusite crystals reaching as long as 10 cm on the west of the Gerze area (Pan Yusheng 1981). In the southern segment, the Gaoligong Mts. represent one of the three famous metamorphic belts in western Yunnan. Along the fracture zone is distributed the second gigantic ultrabasic rock belt in China after the Yarlung-Zangbo River one. In addition, a spilite-keratophyre series is found. The authors also have reason to suspect that the klippen to the north of the Sogxian-Amdo line might be a mélange belt as well, but this needs to be proved through further work.

4. *The Jinsha River-Red River Deep Fracture Zone*

This fracture zone extends more or less along the Jinsha and Red Rivers in the form of a reversed S-shaped arc. It is divisible into the northern and the southern segments. The northern segment, seated to the north of the Shigu area, is known as the Jinsha River Deep Fracture Belt which runs principally along the Jinsha and Tongtian Rivers, passes westwards through the Fenghuo Mt. Pass and Yuejitai County, and is cut by the Ruoqiang-Lazhulong Deep Fracture in the vicinity of the Lazhulong area, thus forming a dividing boundary between the Karakorum-Tanggula-Sanjiang and the Songpan-Garze Fold Systems. The southern segment runs along the Cangshan Mt., the Ailao Mt. and the Red River, and is split into the eastern branch known as the Red River Deep Fracture and the western branch known as the Ailao Mt.-Tengtiao River Deep Fracture. They make up the boundary between the Yangtze Paraplatform and the South China Fold System on one hand and the Sanjiang Fold System on the other. The Jinsha River-Red River Deep Fracture Zone continues extending further to the southeast, passing over the border of China and through the territory of

Vietnam, stretching into the Beibu Gulf (Bac Bo Gulf), and reaching the sea domain of the Yinggehai at the southwestern end of the Hainan Island, with a total length of over 4000 km.

The Jinsha River-Red River Deep Fracture Zone represents a fracture zone that has been polycyclically active at least since the Indosinian Cycle, with which the well-known polycyclic high-grade metamorphic belt of the Cangshan-Ailao Mt. is associated. Along the fracture zone, ophiolitic suites (unpublished data of surveys in Sichuan and Yunnan Provinces) are found in the areas of Derong, Deqen and the Yuanjiang River. Wang Kaiyuan et al. have reported of the discovery of glaucophane-schists in a low-grade metamorphic belt lying on the western side of the Ailao Mt. high-grade metamorphic belt, and held that the Ailao Mt. high-grade metamorphic belt and the low-grade metamorphic belt of it (greenschist facies) constitute precisely paired high-temperature and low-pressure, and high-pressure and low-temperature metamorphic belts (Wang Kaiyuan et al. 1981). Geologists from the Sichuan Regional Geological Mapping Party and Zhang Zhimeng and others have found an ophiolitic melange belt in the area of Derong along the western flank of the fracture. More recently, some geologists consider that the Batang Group in the areas of Yushu and Dainkog might also be ophiolitic mélanges (Zhao's pers. comm. 1976). Therefore, the Jinsha River-Red River Deep Fracture Zone must be an important translithospheric fracture zone (a plate suture).

What is to be explained is that the Jinsha River-Red River Deep Fracture Zone might also represent a still older suture, as the southern segment of the fracture zone, the Red River Deep Fracture south of the Diancang Mt., cuts obliquely the tectonic line of the Kam-Yunnan Axis of the Yangtze Cycle. Meanwhile, the Late Proterozoic-Early Cambrian time range for the oldest metamorphic complexes within the Ailao Mt. metamorphic belt demonstrates that the Red River Fracture has been an important deep fracture belt at least since the Late Proterozoic, and that the later Jinsha River-Red River Deep Fracture Zone is nothing but an inheritance and development of the Red River Deep Fracture itself.

5. The Eastern Kunlun-Southern Qinling Deep Fracture Zone

This fracture zone comprises the eastern Kunlun and Maqen-Lueyang Deep Fractures, which are in fact connected with one another to form a single deep fracture belt running approximately along the Qiman Tagh Mt., the southern margin of the Burhan Budai Mt., and the northern margin of the Jishi Mt. The western end of it is cut by the Ruoqiang-Lazhulong Fracture. To the east it extends for over 2500 km, passing through the Lueyang area up to the north of Hanzhong to form a boundary between the Kunlun and Qinling Fold Systems on one hand and the Songpan-Garzê Fold System on the other. Along the fracture zone, from the Burhan Budai Ma. to the Jishi Mt., are distributed Permo-Triassic ophiolite (Qin Deyu, the Geological Bureau of Qinghai Province, unpublished data) and mélanges with a total length of several hundreds of km (Li Chunyu 1978). In the Burhan Budai Mt., lying on the northern side of the fracture zone extends an 800-km-long granitic belt; and in the area around the Kunlun Mt. Pass south of the fracture belt there are clues showing the possible presence of a high-pressure and low-temperature metamorphic belt (Qin Deyu, unpublished data); thus further investigations are obviously necessary.

6. The Northern Margin of Qaidam-Northern Huaiyang Deep Fracture Zone

This contains the deep fractures located in the Danghenanshan area, on the northern margin of Qiadam, in the Shaliu Riversarea, Qinghai Lake-Northern Huaiyang area, and Lintan-Shanyang area. The Qinghai Lake-Northern Huaiyang Deep Fracture starts from the area near the Da Qaidan, passes through the Qinghai Lake, Linxia County, the area south of Tianshui and Shangxian and Nanyang Counties, and goes on further eastwards along the northern piedmont of the Dabie Mt. to reach the area near Lujiang County, where it is cut by the Tancheng-Lujiang Deep Fracture, with a total length of more than 2200 km, constituting a boundary between the Qinling and Qilian (including the Caledonides in northern Qinling) Fold Systems. The Lintan-Shanyang Deep Fracture lies to the south of the Qinghai Lake-Northern Huaiyang Deep Fracture and extends from the area east of Tongren County in Qinghai Province through the Zhashui and Shanyang Counties and up to the Nanyang County area, where it meets with the Qinghai Lake-Northern Huaiyang Deep Fracture, thus forming a dividing boundary between the Variscanids and the Indosinides in the northern Qinling Mts. According to the data obtained by the Regional Geological Mapping Parties of Gansu, Qinghai and Shaanxi Provinces, and to the results of the thematic research conducted by Li Chunyu (1978), the western segment of the fracture zone contains intermittent occurrences of mélanges in the Triassic geosynclinal deposits spreading in a distance of 800 km from Xiangpi Mt. on the southern bank of Qinghai Lake to Fengxian County, while in the eastern segment, in the area of Shanyang, Shangnan and Neixiang Counties, along the southern side of the fracture zone there exists a high-pressure and low-temperature metamorphic belt (the principal minerals are magnesian riebeckite and aluminian grunerite, with minor glaucophane) extending continuously for 180 km. The Northern Qinling Caledonides on the northern side of the fracture zone form, at the same time, a polycyclic granitic belt (Caledonian, Variscan, Indosinian and Yanshanian). The combination of the two segments should represent paired metamorphic belts associated with a plate suture.

7. The Northern Qilian-Northern Muaiyang Deep Fracture Zone

This fracture zone extends along the Northern Qilian-Northern Qinling Fold Belt and comprises the deep fractures distributed in the Longshou Mt., on the northern edge of the northern Qilian Mts., in the northern

Qilian Mts., and on both the northern and southern edges of the Qinling Axis.

The deep fracture on the northern edge of the Qinling Axis has been active ever since the Proterozoic. It extends westerly, passes through the western margin of the Ordos Basin and connects with the Longshou Mt. Deep Fracture lying to the south of Alxa area, creating a fossil suture between the Yangtze Fold Belt and the Sino-Korean Paraplatform.

The Northern Qilian Deep Fracture takes the form of an arc projecting towards the NE and extends from the area southwest of Yumen City, through Jingtai and Haiyuan Counties, and up to the area of Guyuan County, from where it turns to the south and continues further to the area of Baoii City to meet with the deep fracture on the southern margin of the Qinling Axis.

According to the studies made by Xiao Xuchang et al. (1978), Wang Quan and others (1976), there are well-developed Cambrian and Ordovician ophiolitic suites and a glaucophane-schist belt with a total length of over 100 km in the northern Qilian Mts., serving as a reliable sign to show the presence of a translithospheric fracture zone here. As an eastern extension of the particular fracture zone, the deep fracture belt on the southern margin of the Qinling Axis has been subjected to a strong and extensive migmatization on its northern side. The discovery of a high-pressure metamorphic belt with 3-T metamorphosed phengite and C-eclogite in the Xinyang area (Ye Danian et al., 1979) indicates the presence of a high-pressure and low-temperature metamorphic belt.

It is the long-term activity of the above three gigantic deep fracture zones——the Northern Qilian-Northern Huaiyang, the Qaidam Northern Margin-Northern Huaiyang and the Eastern Kunlun-Southern Qinling Deep Fracture Zones——that have made the Qinling Mts. and their westerly extension (varying in position with different geological times) a boundary between northern and southern China. As was stated earlier (Huang Jiqing et al. 1977), all these fracture zones tend to exhibit salient features such as being deeper in the west and shallower in the east.

8. The Ertix Deep Fracture Zone

This deep fracture zone is located between the Altay and the Junggar Fold Systems and extends northwesterly. Continuing to the northwest, it enters the territory of the U.S.S.R. where it is split into the northern and the southern branches: the northern one is also known as the Northeastern Deep Fracture, and the southern one as the Ertix Deep Fracture. To the southeast, it runs through Mongolia (known as the Central Mongolian Deep Fracture) to join the Derbur Deep Fracture. The whole fracture zone constitutes a gigantic arcuate fracture projecting towards the south, with a total length of more than 4000 km. It forms a boundary between different lithofacies zones in China. Although it has been subjected to compression, fracturing and high-grade metamorphism, no ophiolitic suites have so far been discovered. Meanwhile, in the territories of the U.S.S.R. and of the

Mongolian People's Republic large-scale ophiolitic suites are found which form a boundary between the North Mongolian Xingkai-Caledonides and the South Mongolian Variscides. The particular fracture zone represents one of the deep fracture zones significant to the development of the Central Asian-Mongolian Geosynclinal Region.

9. *The Xar Moron Deep Fracture Zone*

This fracture zone extends along the Xar Moron River in western Liaoning. To the west, it enters the territory of the Mongolian People's Republic through the area south of Erenhot, while to the east it emerges in the vicinity of Changchun City via the southern' end of the Songliao Plain, with a total length of over 1100 km. It is an important tectonic boundary showing quite different geological histories on either side. It is, at the same time, a significant biogeographic boundary line, to the south of which the Permo-Carboniferous biota is essentially warm-water Pacific fauna and Cathysian flora, while the biota on the northern side of the boundary is characterized chiefly by cold-water Arctic fauna and Angora flora. In the Ondor Temple area and elsewhere, ophiolitic suites are encountered. In addition to strong dynamic metamorphism and compression fracturing, glaucophane-schists and mélanges have recently been discovered in many places along this fracture zone.

Besides these several deep fracture zones, the deep fracture on the northern edge of the Inner Mongolian Axis and the deep fracture on the northern margin of the Central Tianshan Mts. may also be translithospheric fracture zones that make up a suture between the Central Asian-Mongolian Geosynclinal Region and the Tarim-Sino-Korean Platform (the Chinese Protoplatform). This is still a problem that needs to be solved by further investigations. The Almantai and Karameili deep fractures in Xinjiang tend to run eastwards to connect with the South Mongolian deep fractures and to enter the territory of the U.S.S.R. when extending towards the northwest. Along these fractures are distributed massive ophiolite suites. Some geologists are of the opinion that this region might in fact be the place where the oceanic crust in the Central Asian-Mongolian Geosynclinal Region finally disappeared. Hence, these deep fractures might equally be translithospheric ones. In the southern part of China, Late Proterozoic ophiolitic suites are distributed along the Jiangshan-Shaoxing and Xupu-Sibao Deep Fractures. According to Guo Lingzhi (Guo Lingzhi's report at the Second All-China Congress on Tectonics, 1979) and Qiao Xiufu et al. (1981), they might represent fossil sutures. The studies made by Luo Zhili (1979) showed that the Longmen Mt. Deep Fracture might also be a Late Proterozoic fossil suture. Again, it was reported by Liu Chang'an and others (1979) that along the Mudanjiang Deep Fracture glaucophane-schists have now been discovered in the areas of Yilan and Jiayin Counties. Whether or not they represent a fossil suture is still a problem worthy of further study.

V.2.2 The Most Important Shear Deep Fractures

Within the territory of China there exist a great number of shear fractures, among which the Tancheng-Lujiang Deep Fracture in the east and the Altun-Beishan Mts. and Yulong-Longmen Mts. Deep Fractures in the west are the largest and most important. Some of them may probably be relicts of transform faults.

1. *The Altun-Beishan Mts. Deep Fracture Zone*

This is a deep fracture zone extending from the Altun Mts. to Mongolia through the Beishan Mt., 100—250 km wide and over 2000 km long, and is represented by a series of NE- and ENE-trending sinistral shear fractures. The major ones are the Qiemohe, Ruoqiang-Lazhulong, Xingxingxia, Liuyuan and Altun Deep Fractures, of which the most attractive is the Altun Deep Fracture that extends for 1500 km and constitutes the northwestern edge of the Qinghai-Tibet Plateau known as the Roof of the World.

The Altun-Beishan Mts. Deep Fracture Zone represents a set of deep fractures which have been active ever since the Paleozoic. In its western segment, the Ruoqiang-Lazhulong deep fracture cuts through the Kunlun Fold System in the middle; while in its eastern segment, the Altun Deep Fracture cuts the Qilian Mts. Fold System at its western end. The Xingxingxia Deep Fracture intersects obliquely the Tianshan Mts. Fold System and makes the latter change its direction from a northwest trend into an approximately eastwest one. Therefore, from both the geological maps and the Landsat photographs we can see, at first glance, that in the western part of China as a whole there lies transversely an eye-catching ENE-trending tectonic belt, known as the Altun-Beishan Tectonic Belt on the background of the general NW-trending tectonic setting. The spatial arrangement of the various folded mountain chains after their displacement, especially the occurrence of gigantic lambda-shaped folds and fractures in the Qaidam Basin on the eastern side of the Altun Deep Fracture and in the western segment of the Qilian Mts. have completely proved that this deep fracture zone has been a set of sinistral shear fractures since the Cenozoic.

2. *The Yulong-Longmen Mts. Deep Fracture Zone*

This refers to a set of NE-trending fractures that extend from the Longmen Mt. to the Yulong Mt. areas and finds clear expression in the geological and geomorphological aspects, as well as in the geophysical anomalies. This is a set of protracted and polycyclic deep fractures. During the Paleozoic and the Early Mesozoic (Triassic period), it served as a boundary between the stable region essentially of platforms in eastern China and the mobile region mainly of geosynclines in western China. In the Late Mesozoic, particularly since the Cenozoic, it used to represent a southeastern boundary of the largest fold-uplift region——the Qinghai-Tibet Plateau in western China.

The Yulong-Longmen Deep Fracture Zone represents a set of shear-compression fractures with compression as the main force. In the Longmen Mt. area not only nappe structures are developed, but also some obliquely arranged flattened rhombic structures which are shown clearly on both the geological maps and Landsat photographs. The rhombs, consisting of the Pengguan complexes in the Pengguan area and the synclinorium composed of Devonian and Carboniferous sequences in the Tangwangzhai area, are the most typical representatives. In the areas of Lijiang and Heqing there is also a series of nappe structures, of which the most typical is that in the vicinity of the Heqing Town where the Triassic Beiya Fromation has been overthrust upon the Tertiary Lijiang breccia. The particular fracture zone extends from Sanying up to the north of Heqing for about 100 km.

The direction of shearing of the Yulong-Longmen Fracture Zone in Early Mesozoic (Indosinian stage) was different from that in the Himalayan stage. The Indosinian Cycle was characterized by sinistral shearing as expressed by a set of drag arcuate fold-fractures, such as the Wudu, Wenxian and Pingwu Fractures on the western side of this fracture zone, while the Himalayan Cycle is characterized by dextral shearing, which can be seen clearly from the Landsat photographs and the geological maps, with the Lijiang and Heqing areas providing the most striking evidence. The Landsat photographs have also demonstrated clearly that the Longmen Mt. Deep Fracture extends towards the northeast after passing through the Hanzhong area, and then enters the Qinling Mts. to join with the Lintan-Shanyang Deep Fracture in the vicinity of Zhen'an County.

The sinistral Altun Deep Fracture Zone forms the northwestern boundary for the Qinghai-Tibet Plateau, while the dextral Yulong-Longmen Mts. Deep Fracture Zone forms the southeastern boundary for the Qinghai-Tibet Plateau. Taking the piedmont huge nappe of the Himalayas and the intensive thrusts and overturnings at the piedmont of the Qilian Mts. into consideration, we may see a clear general tectonic picture showing a compression towards NEN by the Indian Plate, as well as a compression from SW to NE by the Qinghai-Tibet Plateau as a whole.

3. *The Tancheng-Lujiang Fracture Zone*

This fracture zone was first discovered by the Division of Airborne Geophysics and Aerogeology under the Ministry of Geology and Mineral Resources in 1957 (unpublished data of the Division of Airborne Geophysics and Aerogeology under the Ministry of Geology and Mineral Resources). Since then, it has been further studied by the relevant provincial geological organizations, such as the former Beijing College of Geology, the Hefei Technological University, the Institute of Geology of Academia Sinica and the State Seismological Bureau. In addition, Huang Jiqing, Zhang Wenyou and others had many a time mentioned it in their previous papers. More recently, Xu Jiawei (1978) delivered a monograph elaborating on the horizontal displacement of this particular fracture zone.

At present there is still great disagreement among Chinese geologists about the time of formation and the distance of horizontal displacement of this fracture. Nevertheless, they all agree that this is a very important deep fracture zone, of great significance to the development of the geotectonics in China and eastern Asia. During the Mesozoic Era there occurred sinistral shearing to a certain extent along the fracture belt, while during the Cenozoic Era it was characterized by a dextral shearing. Such a variation in the direction of shearing reflects the transformation of the direction of relative movement between the Pacific Ocean and the Asian continent.

The southern end of this fracture zone was previously drawn as far as to the vicinity of Guangji County, Hubei Province. Whether or not it crosses the Changjiang (Yangtze) River and extends further southwards remains a major problem in Chinese geology. With the aid of Landsat photographs and in combination with the analysis of geological data available, it has been found that the Tancheng-Lujiang Fracture Zone is split into two branches to the south of Caohu Lake. Its western branch goes through Guangji, Hengyang, Ningyuan and Pingnan Counties up to the Shiwan Mts, then passes over the border line and enters the territory of Vietnam in the form of a 40—50-km-wide fracture zone restricted by two major fractures; while its eastern branch passes by Anqing City, Ahui Province, crossing the Changjiang (Yangtze) River and Poyang Lake whence it goes essentially along the Ganjiang River (as suggested by Guo Wenkui in the early 1960's) till it meets the Wuchuan-Sihui Fracture. In addition, it has been discovered that another fracture, known previously as the Hepu-Rongxian (now named the Hepu-Yangxiu) Fracture Zone, within the territory of Guangxi Zhuang Autonomous Region, passes through Cangwu, Chenxian and Chaling Counties and extends to Yongxin County and still further to the northeast with a total length of more than 1000 km. Whether it represents another branch of the Tancheng-Lujiang Fracture Zone remains to be solved by further studies.

The northern extension of the Tancheng-Lujiang Fracture Zone was originally thought to be divided into two branches in eastern Liaoning after overpassing the Bohai Sea. The western branch is known as the Yilan-Yitong Deep Fracture, the eastern as the Fushun-Mishan Deep Fracture. It is now observed from the satellite images that a gigantic fracture zone, which starts from Yingkou City and passes through the cities of Shenyang, Changchun and Harbin up to the Sea of Okhotsk within the territory of the U. S. S. R., is shown clearly in the Geological Map of the Far East of the U. S. S. R., and may represent a third branch of the northern extension of the Tancheng-Lujiang Fracture Zone.

Formerly, only a small segment of the well-known Mid-Asian Talass-Fergana Deep Shear Fracture Zone was outlined within the territory of China, but the satellite images show that it passes through the northwestern part of Kashi and traverses the Tarim Basin and the western Kunlun up to the vicinity of the Bangong Lake in western Tibet, and controls the Jurassic basin in front of the western Kunlun. Thus, we call the China portion of this fracture belt as the Kashi-Bangong Lake Deep Fracture Zone.

V.2.3 Crustal Fractures

Crustal fractures are the most numerous among the various kinds of deep fractures. In the 1:4,000,000-scale *Tectonic Map of China,* we have emphatically demonstrated some major crustal fractures occurring in eastern China. For western China, because there is even a greater number of different kinds of more complicated deep fractures, the emphassis is laid only on the lithospheric and translithospheric fractures.

The crustal fractures within China can be classified into two distinct categories: the sialic fractures, represented by the Southeastern Coastal Deep Fracture system that controls the distribution of Mesozoic volcanic rocks and granites in southeastern China, and the simatic fractures that controlled the origing and development of the down-faulted basins in North China, northern Jiangsu and Jianghan region.

1. *The Sialic Fractures*

Such fractures may well be exemplified by the Southeastern Coastal Deep Fracture System. Formerly, only the most evident and most important deep fractures such as the Changle-Xiamen, Lishui-Haifeng, Shaowu-Heyuan, and Wuchuan-Sihui Fracture Zones along the southeastern coast of the country were shown on the tectonic map. Now, however, through the analysis of the satellite images and in combination with the study of both the geological and geophysical data available, we have not only discovered more fracture zones, but have also collated and stipulated the extension and position for some of the originally recognized deep fracture zones (e.g. the Shaowu-Heyuan Deep Fracture Zone).

The Southeastern Coastal Deep Fracture System is characterized essentially by NE- and NEN-trends, but there are also not a few ENE-trending or NW-trending fracture zones. Altogether they form a complicated fracture system, controlling the Mesozoic magmatism and metallogenesis. The Mesozoic Era, especially the Yanshanian Cycle, served as a major period for their extensive activities. but there are fractures that had already been formed in the Paleozoic Era.

a) *The NE-trending deep fractures,* which include the Wuchuan-Xiaoshan, Enping-Longquan, Zhongshan-Wenshou, and Putian-Lufeng Fracture Zones. It should be noted here that the Heyuan Deep Fracture in Guangdong Province was formerly considered as connected with the Shaowu Fracture in northern Fujian, as a single Shaowu-Heyuan Fracture Zone. Yet in the course of our present study, we found that the Heyuan Fracture and the Shaowu Fracture are actually two separate gigantic crustal fractures. Both the Landsat photographs and the 1:200,000-scale geological map clearly demonstrate that the Heyuan Fracture starts in the southwest from the Hailing Island at the coast of Yangjiang area, and extends northeast through Zhongshan and Heyuan up to Longchuan County from where, instead of turning to the Xunwu-Huichang line as was thought previously, it proceeds towards the northeast and passes through western Longyan, eastern Nanping and Wenshou district, and reaches Taizhou Gulf in Zhejiang Province, constituting a 20—30-km-wide and over 1000-km-long gigantic fracture zone restricted by two major fractures. According to the localities where it starts and ends, we call this fracture zone the Zhongshan-Wenshou Deep Fracture Zone. The Shaowu Fracture is part of the gigantic Enping-Longquan Deep Fracture Zone which starts from Enping area and extends through Guangzhou City, the Fogang granite terrain, and Longnan and Ruijin Counties, up to Longquan County or still farther. It is also a fracture zone with a total length of over 1000 km and is roughly paralleled to the Zhongshan-Wenzhou Deep Fracture Zone. The Xunwu-Huichang Fracture which was originally thought to connect the Heyuan with the Shaowu Fracture seems to be also a remarkable fracture zone that extends in an approximately N-S direction and belongs to one and the same set of fractures as the Yihuang-Ningdu and Nancheng-Guangchang Fractures

The Wuchuan-Sihui Deep Fracture is one of the important fracture zones along the southeast coast of China that had been recognized at a quite early date. The geologists of Guangdong Province hold that this deep fracture was formed immediately after the Caledonian Movement and later subjected to polycyclic tectonic activities for a long period, thus forming a complicated squeezed and crushed compression fracture zone as well as a magmatic zone. Yet its northern extension has long been uncertain. Now through a comprehensive analysis of the Landsat photographs and latest geological data, we have found that it joins the Ganjiang Fracture via Shaoguan to form a branch of the Tancheng-Lujiang Fracture Zone in one direction, while in the other it constitutes another gigantic NE-trending fracture zone, passing through Yingde, Nanxiong, Ganzhou, Nanfeng and Shangrao Counties up to the Hangzhou Gulf. It is parallel to the Enping-Longquan and Zhongshan-Wenzhou fractures and has now been renamed the Wuchuan-Xiaoshan Deep Fracture Zone.

The Putian-Lufeng Deep Fracture Zone is located in the coastal areas of Fujian and Guangdong Provinces and also represents a newly defined fracture zone. It starts from the seashore of the Lufeng area in the southwest and extends northeast through the Shantou, Zhangzhou and Quanzhou areas up to the mouth of the Minjiang River. A granitic belt is observed just along this fracture zone. On both the geological maps and Landsat photographs the fracture structures are shown very clearly.

b) *The NEN-trending fractures*, which include the Changle-Xiamen. Ninghai-Zhangzhou and Lishui-Haifeng fracture zones. The Changle-Xiamen Fracture Zone starts in the south from the Dongshan Island and extends northerly along the coastal areas of Fujian and Zhejiang Provinces for over 800 km, as represented by a gravity-anomaly gradient belt in the gravimetric map. On the geological map it is marked as a tectono-magmatic-metamorphic belt. The metamorphic rocks distributed along this fracture zone were taken previously as Archean gneisses. In recent years, however, the regional geological surveys show them to be Mesozoic sequences with granite intrusions that have been subjected to severe dynamometamorphism.

The Lishui-Haifeng Deep Fracture Zone, whose extension within Guangdong Province is named the Lianhua Mt. Fracture, and within Fujian Province the Zhenghe-Dapu Deep Fracture. In the Landsat photographs it appears as a gigantic 20—50 km-wide and 1100-km-long fracture zone bounded by two major faults. Its eastern branch starts in the south from Haifeng and extends north through Jiexi, Dapu, Longxi and Lishui Counties up to the Ningbo area and then enters the Hangzhou Gulf, while its western branch starts in the south from Bao'an and extends north through Huidong, Wuhua, Yongding, Longyan, Nanping, Zhenghe, Lishui, Chengxian Counties up to Cixi County and then also enters the Hangzhou Gulf. The evidence of dynamometamorphism of the rocks along the fracture is extremely remarkable.

c) *The ENE-trending fractures*, which are developed extensitly in the coastal areas from Zhejiang to Guangdong Province and are clearly shown in the Landsat photographs. Among them the Qinzhou-Fuqing Fracture Zone is the most important. This is a huge fracture zone that starts from the mouth of the Minjiang River and extends through Dehua, Longyan, Heping, Fogang and Yunan Counties to join with the Lingshan Fracture when entering the territory of Guangxi Zhuang Autonomous Region, just controlling the NEN-trending granitic belt distributed along the above-mentioned areas. It is expressed in the gravimetric map as a gravityanomaly belt.

d) *The NW-trending fractures,* the most important menbers of which are the Lianshan-Hong Kong and the Ji'an-Dongshan Fractures that control the distribution of the NW-trending granites.

In addition, geological and geophysical data as well as Landsat photographs tend to show that a very distinct gigantic fracture zone, named for the time being the Pingxiang-Ningbo Fracture Zone, begins from Pingxiang via Zhongshan, Yizhang, Wan'an and Nancheng Counties up to the Ningbo area and extends into the sea. Most of widespread Mesozoic granites in South China are located to the southeast of this particular fracture zone.

One of the important salient features of the deep fracture zone in the southeastern coastal area lies in that the mechanical properties of the fractures had been changed for many a time in the course of their development. For instance, the monographic studies and the 1:200,000-scale regional geological surveys conducted in the vicinity of Heyuan County (unpublished data by the Regional Geological Mapping Party of the Geological Bureau of Guangdong Province and the state Seismological Bureau) have proved that the Heyuan Fracture used to be a compression-shear fracture before the deposition of the Late Cretaceous Nanxiong Formation (Yanshanian Cycle). On both sides of the fracture the rocks were strongly compressed and metamorphosed, while approaching the fault the rocks tend to show marked cleavages, thus giving rise to silicified mylonites and granitic mylonites, as well as crushed granitic lenses. The Lantang Group distributed along the fracture forms a lambda-type fold structure, showing the effect of compression accompanied by sinistral shearing. During the deposition of the Late Cretaceous Nanxiong Formation and the Paleogene red beds, this same fracture became a tension fracture. Under these circumstances, the hanging wall of the original thrust slid down rapidly along the fault surface, resulting in the formation of extremely thick down-faulted basin deposits with fault breccia. Later, the rocks were once again subjected to strong compression, thus making the red beds dip steeply towards the fault, forming small folds along the fault. On the margins of the nearby Dengta Basin, the Paleozoic strata have been overthrust on the Cretaceous red beds. In the early Neogene, tensile stress again prevailed along the fracture, as expressed by the down-sliding of the hanging wall to provide conditions for new deposition, as well as for the eruption of basalt along the fracture. Still later, compression shear once again became dominant, bringing about the tilting and folding of the red beds.

2) *The Simatic Fractures*

Such fractures can be exemplified by the Lower Liaohe-North China tensional fracture system (mainly after the upublished data of the former Ministry of Geology and the Ministry of Petroleum Industry) which includes a series of fracture zones, such as the Zhuoxian-Shijiazhuang, Xingtai-Anyang, Cangdong, and Liaocheng-Lankao Fracture Zones. The Zhuoxian-Shijiazhuang and Xingtai-Anyang fractures are seated on the western margin of the North China Faulted Depression and control the western boundary of the Central-Hebei Depression. This is clearly shown by the gravity, magnetic, electric sounding and seismic survey data. These fractures dip to the east, representing a normal step fault with a vertical throw of 3000 m or more.

The Cangdong Fracture, starting in the south from Daming County and extending northerly through Linqing, Deshou and Cangzhou areas up to the vicinity of Tianjin City, constitutes a boundary between the Cangxian Uplift and the Huanghua Depression. The fracture itself also represents an easterly dipping normal step fault with a vertical throw of about 6000 m. On the eastern side of the fracture there are very thick Paleogene deposits, while on the western side the Paleogene sequences are absent, the Paleozoic rocks being directly overlain by the Neogene deposits.

The Liaocheng-Lankao Fracture, also a normal step fault, forms a boundary between the western Shangdong Uplift and the Dongming Depression. Within the uplift on the eastern side of the fracture the Upper Tertiary overlies directly the Paleozoic, while in the depression on the western side of the fracture there are well-developed the Upper and the Lower Tertiary with a maximum thickness reaching over 9000 m.

The common salient features of these fractures lie in that (Xu Zhiqin, unpublished data) (Fig. 23):

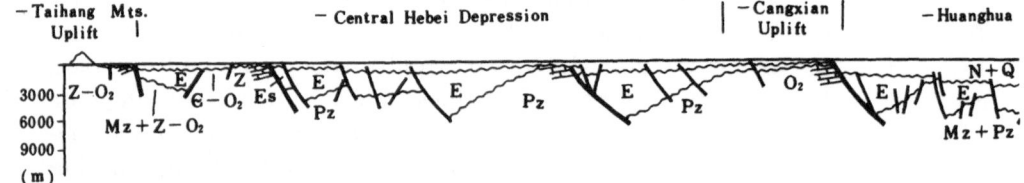

Fig. 23. Simplified profile of the structures of the North China Faulted Depression.

a) They are all tensional faults, representing step faults in a section, and are accompanied by a great number of Y-shaped second-order faults, while in a plane they take the form of Z-shaped broken lines accompanied by en echelon second-order tensional fractures obliquely crossing the major fractures, showing that the fractures are dextral tension-shear ones;

b) The fractures and the basins were developed simultaneously, thus forming winnowing-fan-shaped depression boundaries to control the distribution, lithofacies and thickness variations of the Tertiary strata. Both the maximum vertical throw and horizontal displacement may reach 10 km or more;

c) Along the fractures are distributed paternoster basic volcanics, composed essentially of andesitic basalt, olivine basalt and trachyandesitic basalt, as well as minor trachytes. Petrochemically, they tend to be between alkaline basalts and tholeiites;

d) The fractures are clearly shown by geophysical data. Seismic sounding data have demonstrated that they cut through the Moho;

e) Seismic sounding data have also indicated that the North China Tensional Fracture System occurs on a deep-seated tectonic background with a rift valley-type crust-mantle structure (Teng Jiwen 1974; this paper suggests two models for the crust-mantle structure. In model II, the dotted line indicates that the model is a rift-valley model. In order to make things clear the writer once discussed the problem with Dr. Teng Jiwen).

The North China Faulted Depression is in fact a rift valley-type basin formed by back-arc spreading of the folded mountain system on the eastern margin of the Asian continent (Since the Late Mesozoic time, including Late Yanshanian and Himalayan).

V.3 THE RELATIONSHIP BETWEEN TECTONIC FRAME-WORK OF THE FRACTURES AND DEEP-SEATED STRUCTURES OF CHINA[12]

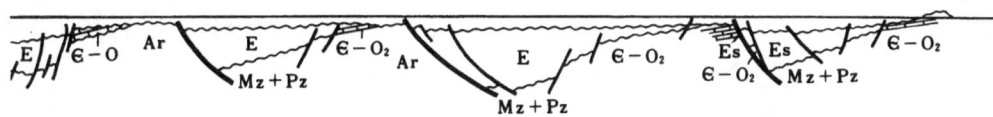

(After the Ministry of Petroleum Industry, unpublished data)

In the process of the crust-mantle structural development, as geological history prograssed, the tectonic pattern of the surface and deep-seated layers of the Earth and the manner of motion of matter changed continuonsly. On the basis of the preliminary analysis of the geological, geophysical and seismogeological data, we have pointed out that the layer-block structure of the continental crust within the territory of China has a very clear expression, and the boundaries between the various crustal blocks commonly represent protractedly developed deep fracture zones. The tectonic aspect of the Moho (Fig. 24) has resulted from the joint operation of the Marginal Pacific and Tethys-Himalayan tectonic domains since the Indo-sinian, especially the Himalayan Cycles. In compiling the 1:4,000,000-scale *Tectonic Map of China*, the comprehensive analysis of the geological and geophysical data and of the Landsat photographs has helped us to deepen our understanding of the particular problem (Figs. 24—30).

1. For the topographic contour lines, the gravity isolates (Fig. 25), the Moho isobath lines and the tectonic lines since the Mesozoic Era in eastern China, the NE-and NEN-trends are predominant, which undoubtedly reflects the result of the motion of matter from the crust and mantle in the Marginal Pacific Tectonic Domain since Mesozoic times. A clear tectonic line is encountered (the Greater Hinggan Mts.-Taihang Mts.-Wu ing Mt. line), representing a gravity anomaly gradient belt, a belt of sharp change in thickness of crust and also a fracture belt termed the Greater Hinggan Mts.-Taihang Mts.-Wuling Mt. Fracture Zone. It is this very fracture zone that splits eastern China into two major parts as represented respectively by their own history of development and quite different salient features in the structure of the crust and mantle. So we hold that it is an important deep-seated structure whose importance (especially since the Late Yanshanian and Himalayan Cycles) is by no means less than the well-known Tancheng-Lujiang Fracture Zone.

---

12 In the course of preparing the present chapter the writer referred to a great deal of the relevant unpublished data obtained by th Institute of Geophysics of Academia Sinica, the State Seismological Bureau and the Institute of Geophysical and Geochemical Exploration under the Ministry of Geology.

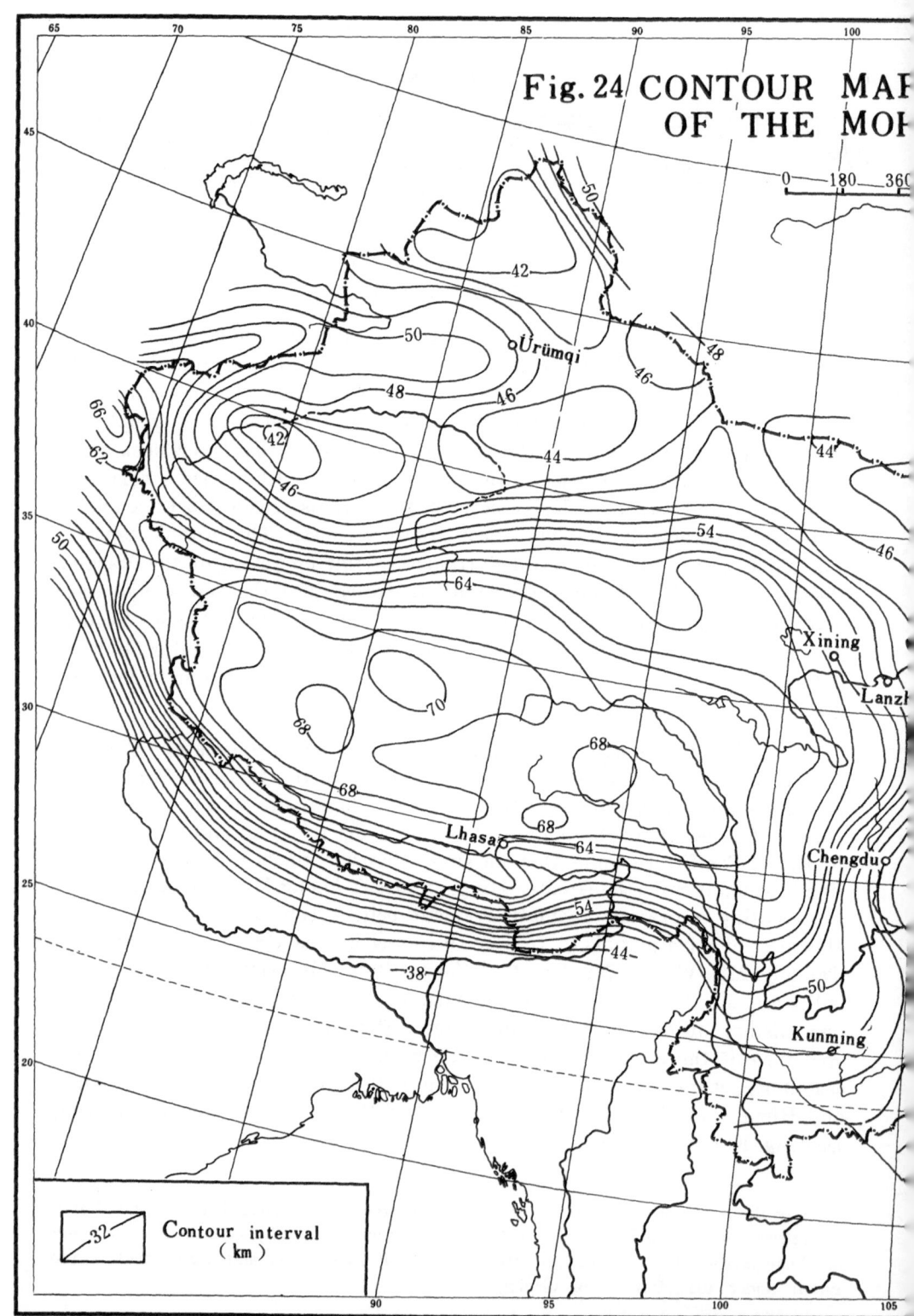

Fig. 24 CONTOUR MAP
OF THE MOH

Contour interval
(km)

(after the Institute of Geophysical Exploration, State Bureau of Geology, unpublished data)

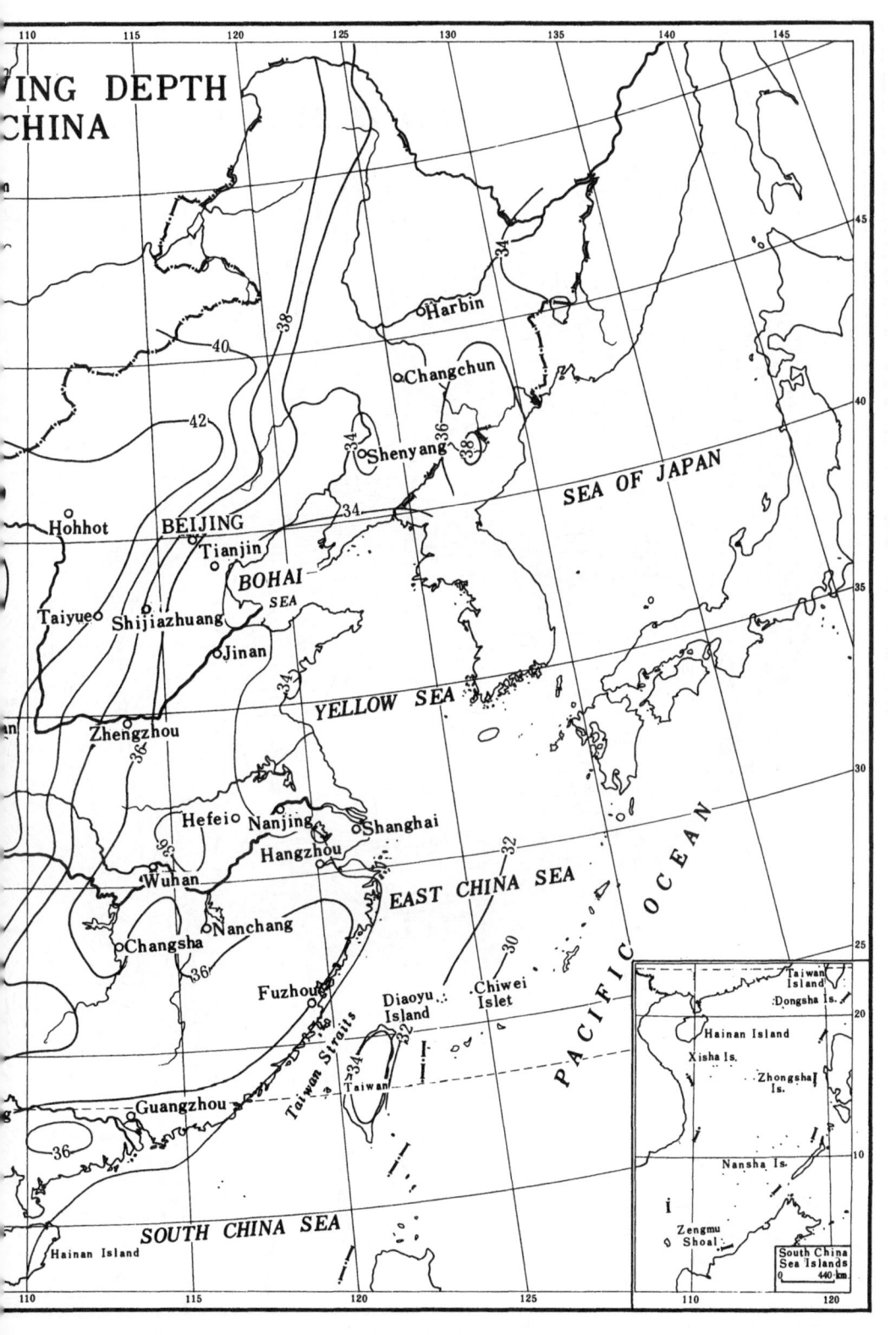

ING DEPTH
CHINA

Harbin

Changchun

Shenyang

SEA OF JAPAN

Hohhot
BEIJING
Tianjin
BOHAI
SEA
Taiyue
Shijiazhuang
Jinan

Zhengzhou

YELLOW SEA

Hefei
Nanjing
Shanghai
Hangzhou
Wuhan
EAST CHINA SEA
Nanchang
Changsha
Fuzhou
Taiwan Straits
Diaoyu
Island
Chiwei
Islet
Guangzhou
Taiwan

PACIFIC OCEAN

SOUTH CHINA SEA

Hainan Island

Taiwan
Island
Dongsha Is.

Hainan Island
Xisha Is.
Zhongsha
Is.

Nansha Is.

Zengmu
Shoal

South China
Sea Islands
0   440 km

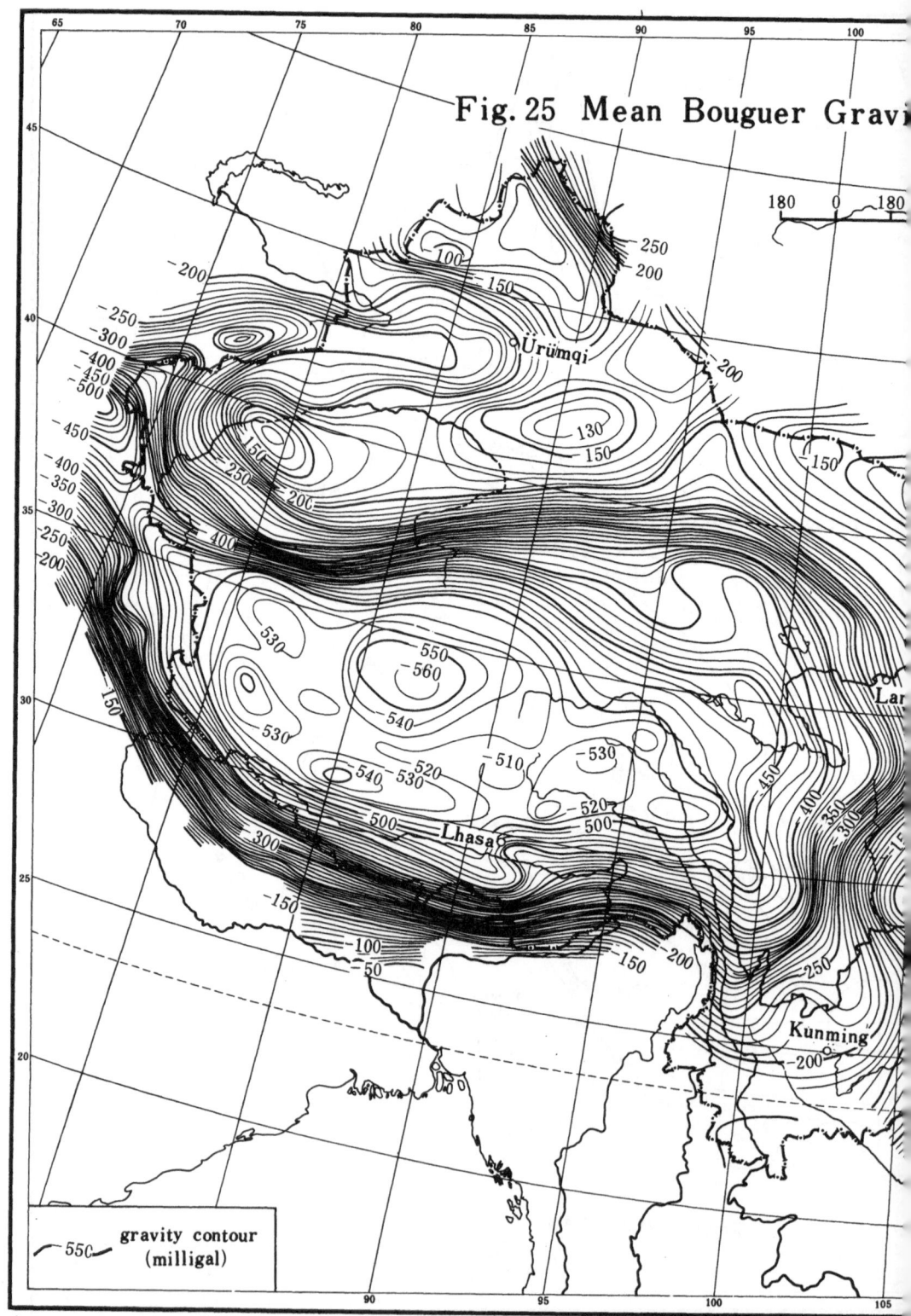

Fig. 25 Mean Bouguer Gravity

gravity contour
(milligal)

(after the Institute of Geophysical Exploration, State Bureau of Geology, unpublished data)

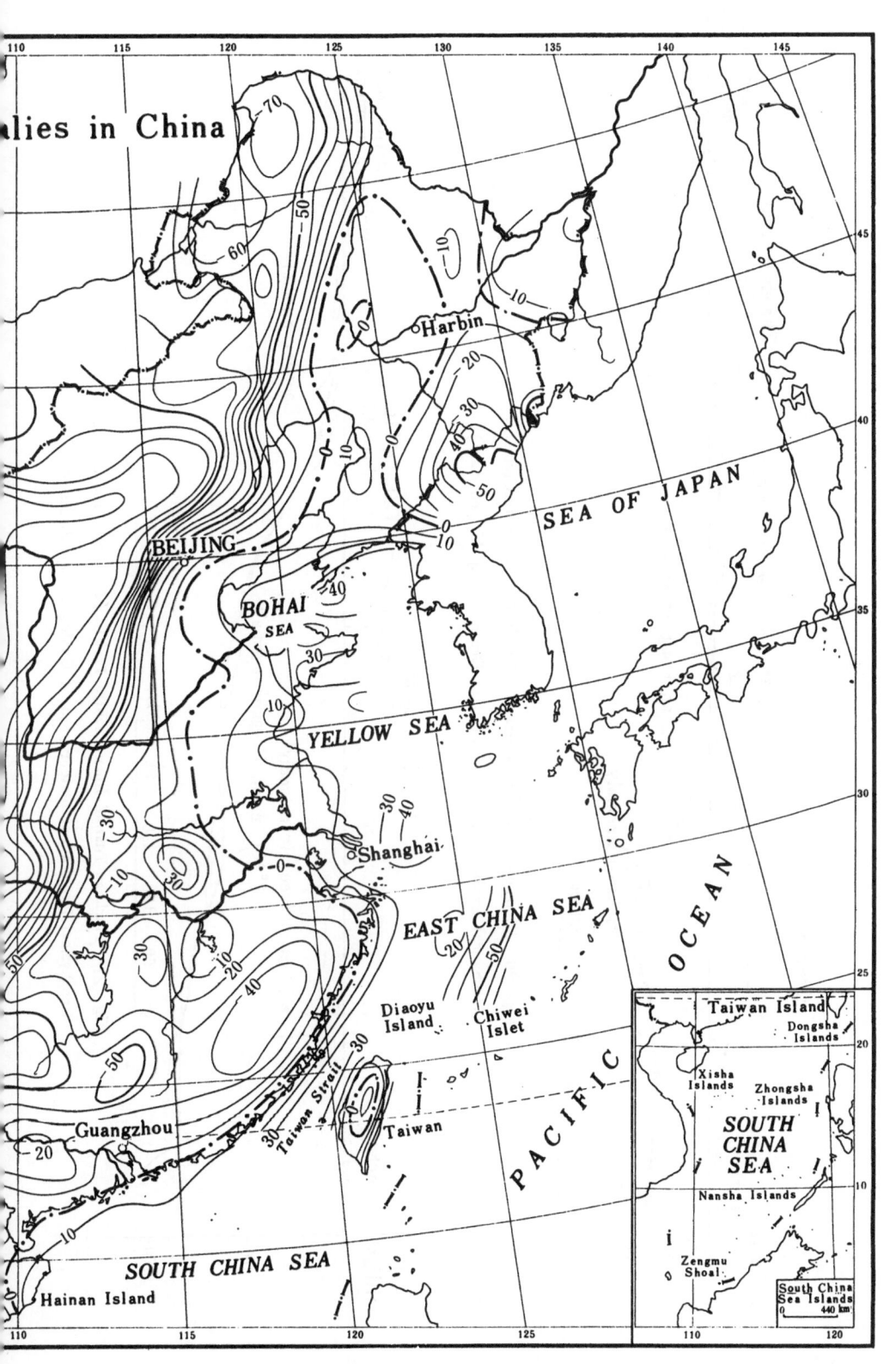

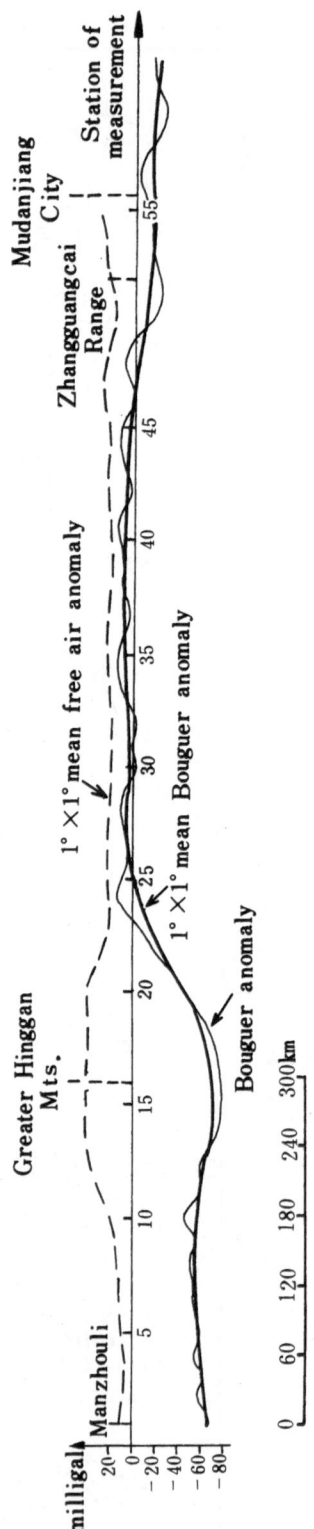

Fig. 26 Manzhouli-Mudanjiang Gravity Profile (after the Institute of Geophysical Exploration, State Bureau of Geology, unpublished data)

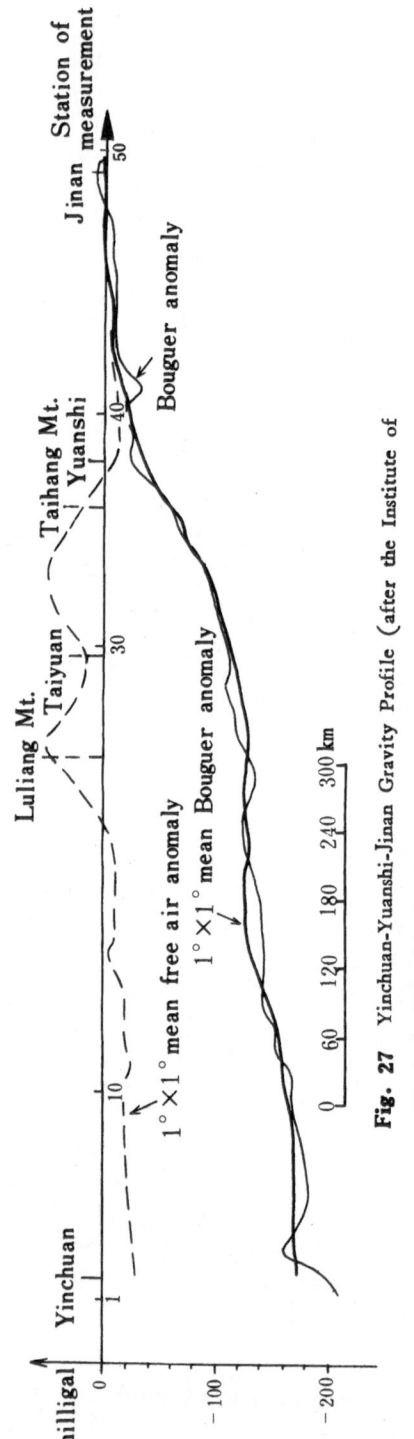

Fig. 27 Yinchuan-Yuanshi-Jinan Gravity Profile (after the Institute of Geophysical Exploration, State Bureau of Geology, unpublished data)

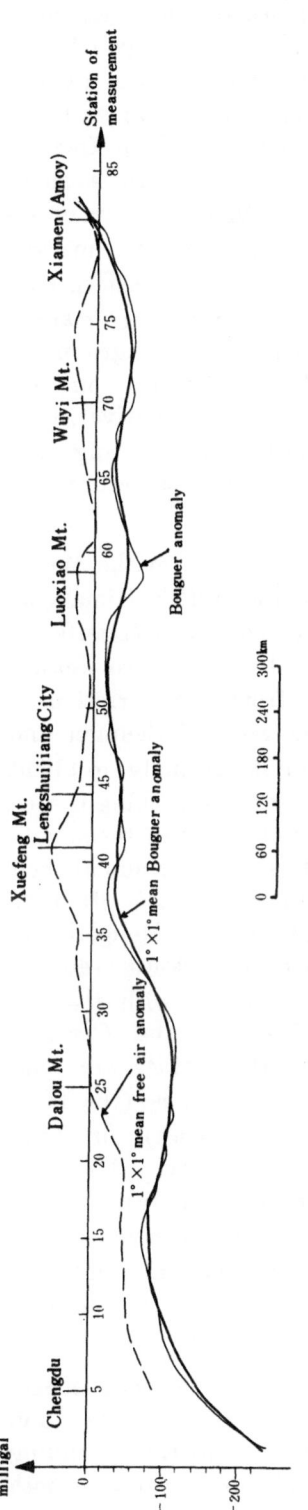

Fig. 28 Chengdu-Xiamen Gravity Profile (after the Institute of Geophysical Exploration, State Bureau of Geology, unpublished data)

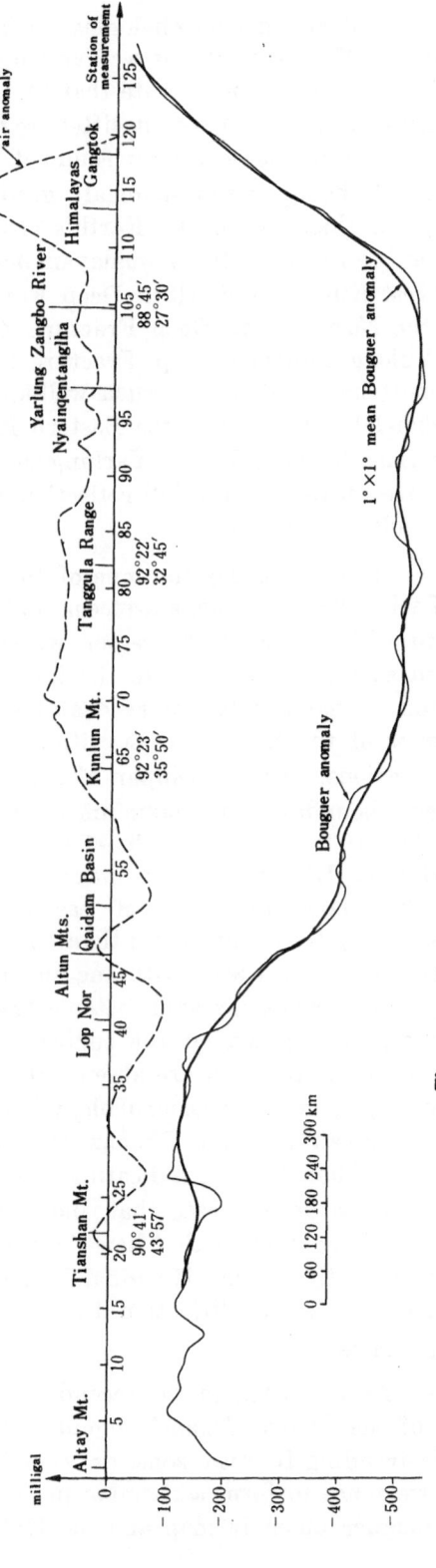

Fig. 29 Altay-Himalayas Gravity Profile (after the Institute of Geophysical Exploration, State Bureau of Geology, unpublished data)

2. All the geomorphological, gravimetric and crustal thickness maps (Figs. 24, 25, 29 and 30) as well as the geological and tectonic maps of western China clearly demonstrate that the Qinghai-Tibet Plateau, as a gigantic structure and the highest uplifted region on Earth, is a region with the thickest continental crust surrounded by several remarkable deep fracture zones (all being gravity-anomaly gradient belts and belts with a sharp change in thickness of the Earth's crust). On the southern margin of the Plateau there occur the Himalayan nappes; on the northwestern margin, the West Kunlun and Altun Deep Fracture Zones; on the northeastern margin, the North Qilian Deep Fracture Zone; and on the southeastern margin, the Yulong-Longmen Deep Fracture Zone. On this general background of the crust-mantle structure, such well-known deep fracture zones on the Qinghai-Tibet Plateau as the Eastern Kunlun, Jinsha River-Red River, Bangong Lake-Nujiang River, Yarlung Zngbo River, and Daofu-Kangding Deep Fractures have found a full reflection in both the deep-seated structures and the Landsat photographs.

3. If we consider that one of these two regions belongs to the Marginal Pacific Fracture Maga-systeem, and the other to the Tethys-Himalayan Fracture Mega-system, the region where the Sichuan and Ordos Massifs are located should be assigned to the intermediate area between the two tectonic domains. That can be observed at the first glance from the variation in thickness of the Earth's crust within China. To the west of Sichuan and Ordos regions, with the Liupan Mt.-Longmen Mt. gravity-anomaly gradient belt as a boundary, the maximum thickness of the crust within China is seen to be 50—70 km in the Qinghai-Tibet Plateau, while to the east of them, with the Greater Hinggan Mts.-Taihang Mts.-Wuling Mt. gravity gradient belt as a boundary, the crust of eastern China together with its eff-shore areas, clearly thins out to less than 40 km. Sichuan, Ordos and Mongolia are in fact located in a zone with one and the same order of thickness (40—50 km). This very deep-seated tectonic background is consistent with the tectonic aspect exposed on the surface. There the Mesozoic and Cenozoic gigantic deep fractures are absent, the structures on the surface are relatively simple, and the geomorphology is characterized by plateaus (the Mongolian Plateau and the Shaanxi-Gansu-Ningxia Loess Plateau) or hilly basins (Sichuan Basin), indicating a relative tectonic stability. All these facts seem to demonstrate that, like the Mongolian Plateau, central Sichuan and Ordos are regions that have been subjected to weaker operation of the Tethys-Himallyan and Marginal-Pacific tectonism, and they all represent regions with less modification by the motion of the Pacific and Indian-Ocean plates.

4. As we stated in the second section of Chapter V, in the western segment of the Central-Asian-Mongolian Geosynclinal Region, the NW- and WNW-trending fracture zones are combined with the NE- and ENE-trending fracture zones to form a rhombic pattern, which is reflected clearly in both the Bouguer anomaly map and the Moho isobath diagram of China.

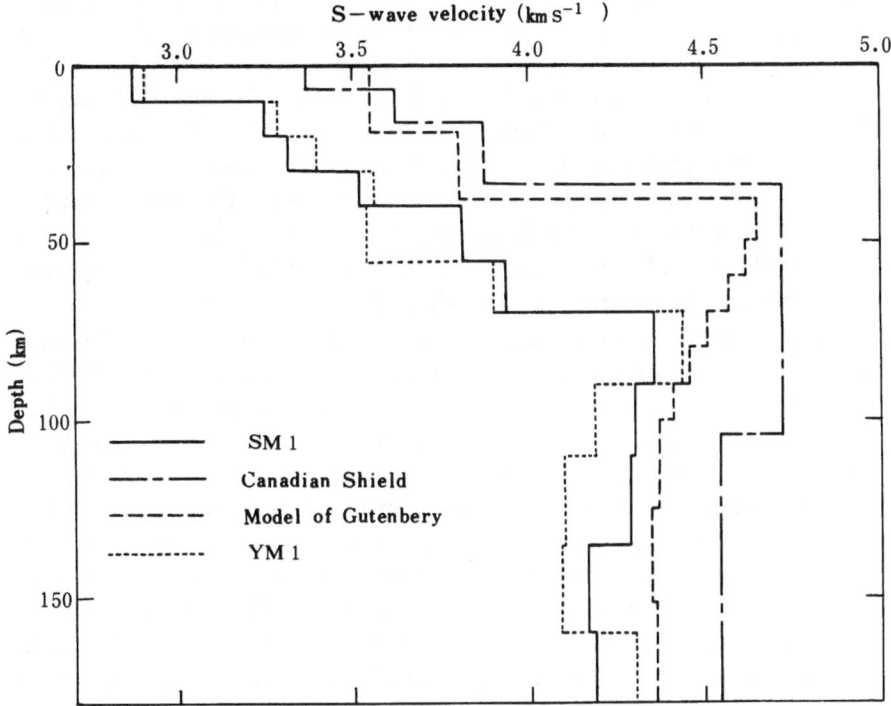

Fig. 30 Model for the Crust-Upper Mantle of the Qinghai-Tibet Plateau

(after Deng Daliang, unpublished data)

SMI and YMI-Model for the Qinghai-Tibet Plateau

5. According to Deng Daliang et al. (unpublished data) within the territory of China the seismic-wave velocity for the crust and the upper mantle is commonly lower, especially in the Qinghai-Tibet Plateau (Fig. 30). This shows that the substances of the crust and upper mantle of China possess a higher activity, while the crust and upper mantle of the Qinghai-Tibet Plateau are in a state of partial melting, thus rendering the propagation of the SM-waves ineffective (Wyllie 1971). We think that this furnishes the deep-seated tectonic background for the modern crust within the territory of China with a higher activity; and with this very activity of the substances of the crust and upper mantle, the neotectonic movements in our country are so intense and active that China has become one of the countries in the world with relatively serious earthquake hazard. This must also be the reason why the Landsat photographs covering the territory of China show so clearly the various tectonic phenomena, particularly the complicated lineaments.

6. The general situation of the crust and upper mantle of China lies in the fact that eastern China, especially the epicontinental basins (such as the North China-Bohai Sea Basin and the northern Jiangsu-Huanghai Sea Basin) and the marginal sea basins (such as the South China Sea Basin),

represents a wide upwarping of the upper mantle, while western China, particularly the Qinghai-Tibet Plateau, represents its downwarping. The former is characterized by an extensional tectonic background; and the latter, by a compressional tectonic background. Such a salient feature is basically consistent with the state of distribution of the various deep fracture zones in our country. The presence of the several tensile fracture zones of the Marginal Pacific Tectonic Domain is also consistent with the wide upwarping of the upper mantle, while the downwarping regions in the west serve as an area of distribution of thrusts and overthrusts, which are expressed more remarkably on the margins of the Qinghai-Tibet Plateau.

7. The present-day structures of the crust and upper mantle of China are essentially products of the combination of the operation of the Pacific Ocean with the Indian Ocean, as well as of the Asian continent. For this reason, the most distinct tectonic aspects that we can see today have been created since the Indosinian, especially since the Himalayan Cycle. However, there is an important salient feature to be mentioned here, which lies in the fact that the new fractures tend to operate usually along the preexisting fractures. That is why certain important fracture zones are always protractedly and· polycyclically active, thus leaving imprints of old structures on the general tectonic pattern of Mesozoic and Cenozoic eras. The most evident case is the presence of several fracture zones within the Kunlun-Qinling Fold Region. The segment from the Qinling Mts. to northern Huaiyang, in particular, constitutes an outstanding geological and geophysical boundary between North and South China, but was cut across by the Mesozoic and Cenozoic NE- and NEN-trending general tectonic framework in the eastern part of China. Zeng Rongsheng (1973) pointed out for certain that the Qinling-northern Huaiyang line is a deep-seated tectonic zone, and that the Lanzhou-Chengdu-Kunming line represents a deep-seated tectonic zone as well. This statement tallies basically with the dividing line between the platforms in eastern China and the geosynclines in western China. The northern margin of the Inner Mongolian Axis, as a boundary between the Inner Mongolian Geosyncline and the Sino-Korean Platform, has actually found its reflection in the deep-seated tectonics, but is less marked than that of the Qinling Mts. area. In addition, the western segment of the Central Mongolian Deep Fracture (from western Mongolia to Ertix) also exhibits a clear gravity-anomaly gradient in the gravimetric map.

Chapter VI

The Geotectonic Evolution of China

The geotectonic evolution of China as summarized in Table 4 is characterized by three most important consolidation stages: (1) the consolidation of the Zhongtiao Cycle at the end of Early Proterozoic (1,700 my), resulting in the formation of the Sino-Korean Paraplatform; (2) the consolidation of the Yangtze Cycle at the end of the Late Proterozoic (700 my) resulting in the formation of the gigantic Chinese Protoplatform; (3) the consolidation of the Variscan Cycle at the end of the Paleozoic, resulting in the formation of the even more extensive Paleo-Eurasia. There are three most important turning points in the geotectonic evolution of China since Late Proterozoic: (1) the disintegration of the Chinese Protoplatform since the Middle Cambrian and the formation of the Paleozoic tectonic framework in China after the Xingkai Cycle; (2) partial fracturing and disintegration of Paleo-Eurasia since the Indosinian Subcycle at the end of the Late Triassic, forming progressively the Meso-Cenozoic tectonic framework of China; and (3) the formation of the modern tectonogeomorphic characteristics of China since the Himalayan Subcycle in consequence of the collision between Eurasia and the Indian subcontinent and of the emergence of the island arcs and marginal seas in the West Pacific (Table 4).

VI.1 THE ARCHEAN AND EO-ALGONKIAN MEGACYCLE (ARCHEAN-EARLY PROTEROZOIC)——GRADUAL FORMATION OF THE SINO-KOREAN PARA-PLATFORM

No definite Archean sequences have been found in most parts of China except in North China. Though the Dagelagebulak Group in the Kuruktag area on the northern margin of the Tarim Basin is assumed to be of Archean age, yet it is still poorly studied. This chapter thus describes only the Sino-Korean Paraplatform.

The Sino-Korean Paraplatform is a large stable region which formed at the earliest date within the territory of China. The process of formation of its basement was roughly divided by us into the Fuping, Wutai and Zhongtiao stages (Huang Jiqing 1977). Recently, Zhu Ying (1979) and others have elaborated in a more comprehensive and systematic way on the development of structure of the basement of the Sino-Korean Paraplatform.

VI.1.1   The Fuping Cycle and Older

In recent years, radiometric determinations, giving ages of 3000 my and older, have been gained by different geological institutions for the Archean metamorphic rocks from eastern Hebei, eastern Liaoning, western Henan and other places which may represent the oldest tectono-magmatic cycle known in China. So far, the Fuping Cycle, occurring at the end of the Archean, has provided accurate geological data and a great number of isotopic datings. A pronounced angular unconformity between the Wutai Group of the Early Proterozoic and the Fuping Group of the Archan can be seen in the Wutai Mt. area, while considerable amounts of isotopic data yielding an age of about 2500 my have been obtained for the Qianxi Group in eastern Hebei, for the Anshan Group in eastern Liaoning, for the Taishan Group in western Shandong, and for the Dengfeng and Taihua Groups in western Henan. Ma Xingyuan holds this for a geological period when the Early Archean continental nucleus was rapidly growing, resulting in the formation of the embryonic platform. The Hehuai nucleus mentioned by Zhu Ying (1979) is likely to be the product of tectonism of this geological time, while the northern part of Ordos (the Baotou-Yinchuan magnetic field) might represent another nucleus.

VI.1.2   The Wutai Cycle

The Wutai Cycle is the most important cycle for the formation of the basement of the Sino-Korean Paraplatform. According to Ma Xingyuan et al., the tectonic framework of the Sino-Korean Paraplatform at that time was: the Ordos and Hehuai nuclei used to be two major stable areas which he termed respectively the Ordos and the Hehuai embryonic platforms. Areas such as the Inner Mongolia-northern Hebei-western Liaoning area, the Alxa area, Jiaoliao area, Huaibei area, as well as the Hengshan Mt.-Wutai Mt.-Lüliang Mt.-Pingliang area, all belonged to tectonically mobile belts in which the Wutai Group and the equivalent eugeosynclinal formations were deposited. With the coming of the Wutai Orogeny occurring approximately 2000 my ago, the geosynclines were folded; as a result, the Hehuai and Ordos embryonic platforms were welded together, leading to the formation of the bulk of the Sino-Korean Paraplatform.

VI.1.3   The Zhongtiao Cycle

This was a period during which the formation of the Sino-Korean Paraplatform was complete. Owing to the fact that the greater part of the Sino-Korean Paraplatform was basically formed after the Wutai Cycle, the remaining mobile belts are characterized by miogeosynclinal deposits, with small amounts of intermediate-basic volcanisms taking place merely at individual localities. Ma Xingyuan et al. have distinguished a number of these

Table 4. Subdivision of tectonic cycles and geotectonic development of China

Era	Period	Epoch	Age (m. y.)	\	\	\	Subdivision	Subdivision of tectonic cycles and geotectonic development
Cenozoic	Quaternary	Q	1.5±0.5	Alpine	Himalayan	H	H²	The collision between the Indian subcontinent and Eurasia, the upheaval of the Qinghai-Tibet Plateau and the formation of the marginal seas of the western Pacific
Cenozoic	Tertiary	N₂	9±3	Alpine	Himalayan	H	H¹	
Cenozoic	Tertiary	N₁	25±2	Alpine	Himalayan	H	H¹	
Cenozoic	Tertiary	E₃	37±2	Alpine	Himalayan	H	H¹	
Cenozoic	Tertiary	E₂	58±4	Alpine	Himalayan	H	H¹	
Cenozoic	Tertiary	E₁	67±3	Alpine	Himalayan	H	H¹	
Mesozoic	Cretaceous	K₂	137±5	Alpine	Yanshanian	Y	Y³	Partial destruction and disintegration of Paleo-Eurasia and the beginning of intensive activity in the Marginal-Pacific Tectonic Domain and the Tethys-Himalayan Tectonic Domain
Mesozoic	Cretaceous	K₁		Alpine	Yanshanian	Y	Y²	
Mesozoic	Jurassic	J₃		Alpine	Yanshanian	Y	Y¹	
Mesozoic	Jurassic	J₂	195±5	Alpine	Yanshanian	Y	Y¹	
Mesozoic	Jurassic	J₁		Alpine	Yanshanian	Y	Y¹	
Mesozoic	Triassic	T₃		Alpine	Indo-Sinian	I		
Mesozoic	Triassic	T₂		Alpine	Indo-Sinian	I		

Geological time

The stage of violent activity in the Marginal-Pacific and the Tethys-Himalayan Tectonic Domains

Geological time				Subdivision of tectonic cycles and geotectonic development		
Era	Period	Epoch	Age (m. y.)			
Paleozoic	Triassic	T_1		I		The closing of the Central-Asian-Mongolian Geosynclinal System, etc., the merging of the Siberian Platform with the Sino-Korean–Tarim Platform and the formation of Paleo-Eurasia
	Permian	P_2	230 ± 10	V^4		
		P_1	285 ± 10	V^3	Variscan	
	Carboniferous	C_3		V^2		
		C_2	350 ± 10			
		C_1				
	Devonian	D_3		V^1		The formation of the South China Platform
		D_2				
		D_1	405 ± 10			
	Silurian	S_3		C^2	Caledonian	
		S_2				The beginning of disintegration of the Chinese Protoplatform and the gradual formation of the Kunlun, Qinling, Beishan, Tianshan (in the mid-south) Geosynclinal Systems, etc.
		S_1	440 ± 10			
	Ordovician	O_3		C^1		
		O_2				
		O_1				
	Cambrian	$\in_3$	500 ± 15			
		$\in_2$				

The stage of gradual formation of Paleo-Eurasia and its development

The stage of gradual formation of the Chinese Protoplatform

Era	System / Group	Symbol	Isotopic age (Ma)	Tectonic name	Symbol	Stage of formation
	Sinian	ϵ_1	600±10	Xingkai	X	The formation of the Yangtze and Tarim Platforms which may be merged with the Sino-Korean Platform, etc. to form the Chinese Protoplatform
	Sinian	Z_z	700±20	Yangtze	A	
Proterozoic	Qingbaikou	Z_q	850±50	Yangtze		
	Jixian	Z_j	1100±50	?		
	Nankou	Z_n	1400±50	Wuling	U	
	Changcheng / Hutuo	Z_c	1700±50	Zhongtiao	Z	The formation of the Sino-Korean Paraplatform
	Wutai	Pt_w	1950±50	Wutai	W	
Archean	Fuping and Older	Ar	2500±100	Fuping and Older	F	The beginning of formation of continental nucleus

Notes: (1) The stratigraphic nomenclature employed here is the same as that adopted in the Geological Map of the People's Republic of China Scale 1:4,000,000, (Chinese Academy of Geological Sciences 1976) with new symbols being established for the Nankou System and Wutai Group; (2) Isotopic ages: Those for the Precambrian sequences are based on the data obtained in China, while those for the post-Cambrian sequences are based on the data recommended by the International Commission on Geochronology (1967).

mobile belts including the Jiaoliao, the Inner Mongolia Autonomous Region, northern Hebei, Hutuo, Lüliang, Zhongtiao and Songji areas, which were consolidated by the Zhongtiao Movement taking place at 1700 my, resulting in the final formation of the basement of the Sino-Korean Paraplatform.

VI.2 THE NEO-ALGONKIAN MEGACYCLE (LATE PROTEROZOIC) ——THE FORMATION OF THE CHINESE PROTOPLATFORM

VI.2.1 The Two Different Types of Sedimentation in the Late Proterozoic

In consequence of the formation of the Sino-Korean Paraplatform, the sediments of Late Proterozoic can obviously be classified into two different types; one is unmetamorphosed, as represented by the Jixian section in eastern Hebei, ofter referred to as the North China type; the other consists of slightly metamorphosed strata, as represented by the Kunyang, Sibao, and Banxi Groups and their equivalents in South China, referred to as the Yangtze type.

The North China type, representing the sedimentary cover of the Sino-Korean Paraplatform, has been subdivided into four systems (Changcheng System, Nankou System, Jixian System, and Qingbaikou System) on the basis of the studies of lithofacies, biostratigraphy and isotopic geochronology.

The Yangtze type, constituting the basement of the Yangtze Paraplatform, is geosynclinal. Its upper part consists mainly of miogeocynclinal deposits represented by flysch and flysch-like rocks (Banxi Group) and by clastic and carbonate rocks (Kunyang Group), with a great thickness, while its lower part is essentially eugeosynclinal; for instance, the Sibao Group in northern Guangxi, the Lengjiaxi Group in western Hunan, and the Fanjingshan Group in eastern Guizhou all contain large amounts of spilite and ultrabasic rocks. The Dahongshan Group below the Kunyang Group in eastern Yunnan, consisting of sodium-rich and iron-bearing metamorphosed volcanic rocks, is especially noteworthy.

For a long time, the Banxi, Lengjiaxi, Kunyang, and Dahongshan Groups were taken for Lower Proterozoic. In recent years, as a result of detailed regional geological survey, and of biostratigraphical and geochronological studies, Chinese geologists came to understand that these stratigraphic groups are not of the Late Proterozoic, but are roughly the equivalents of the Jixian section which used to be considered as Sinian (sensu lato), whereas the section in the Yangtze Gorges, consisting of sandstones, tillites and dolomites, occupies a position higher than the Qingbaikou System of the Jixian section and lower than the basal Cambrian, that is to say, the Sinian system established in the Yangtze section is lacking in the Jixian section. So the Yangtze section is now taken by stratigraphers as a standard for the Sinian System. This latter section is fossiliferous, yielding sponge spicules, vermes, and a large number of microfossils were found. In other words, the Yangtze section represents the standard Sinian, and the term Sinian is no longer applicable to the Jixian section, though formerly it was considered as Sinian (sensu lato).

VI.2.2 The Yangtze Orogenic Cycle and the Formation of the Chinese Protoplatform

The concept of the Yangtze Cycle was proposed by Huang Jiqing et al. in 1974 on the basis of new advances in the study of the Upper Proterozoic in China. The Yangtze Cycle consists of two tectonic movements. The older movement is termed the Jinning Movement, about 850 my in age as shown by an unconformity between the Kunyang Group and its overlying Chengjiang Sandstone which represents a molasse formation. The younger movement is termed the Chengjiang Movement. referring to the unconformity between the Chengjiang Sandstone and the Nantuo Tillite, with an isotopic age of about 700 my. In western Hubei and northern Jiangxi, unconformably below the Sinian Liantou Sandstone there occur two molassic formations, the Macaoyuan Formation and Luokedong Formation respectively. These, in turn, rest unconformably on the Shennongjia and Shuangjiaoshan Groups. This latter unconformity possibly represents the first phase of the Yangtze Cycle.

In consequence of the tectonic movements of the Yangtze Cycle, the crust occupying the present Yangtze Valley as well as the region of the Tarim Basin was practically consolidated, thus forming both the Yangtze Paraplatform and the Tarim Platform, with the sediments of the Sinian System forming the sedimentary cover of the platforms.

It should be pointed out that most of the older metamorphic rocks from the major geosynclinal systems in China have been ascribed to the Late Proterozoic, their sedimentary sequences being of Yangtze type. For instance, the Kawabulak Group, bearing *Anulatus regulatus, Kussiella Kussiensis* Ampiificata and *Conophyton* sp., and constituting the median uplift of the Tianshan Mts. was unconformably overlain by the trilobite-bearing Cambrian phosphatic sediments; the Huangyuan Group and the Kesuer Formation containing *Cryptozoon* sp., forming the median uplift of the Qilian Mts., are unconformably overlain by the Middle Cambrian; the Jinshuikou Group, bearing Late Proterozoic stromatolites, and constituting the median uplift of the Kunlun Mts., is unconformably overlain by the Quanji Formation corresponding to the Sinian System; the Kuanping and Taowan groups, making up the Qinling Axis, are unconformably overlain by the Luoquan Tillite. In the meantime, the platform-type sediments basically analogous to the Sinian System of the Yangtze Gorges are discovered not only in the Yangtze Paraplatform and the Tarim Platform, but are also found in Qinling, Qilian, Kunlun, and Tianshan geosynclinal systems, as well as on the northern margin of the Qaidam Basin. These facts show that in the Sinian period, these regions were not in a state of mobile geosyncline, but had changed into a stable platform. In other words, having undergone the Yangtze Orogenic Cycle, a post-Yangtze-Cycle Platform on a grand scale was formed, extending from the Tianshan Mts. and Tarim in the west, eastward via the Qaidam Basin, eastern Kunlun Mts., Qilian Mts., and Qinling Mts. down to the middle and lower reaches of the Yangtze River. Furthermore, it is most likely that this platform was connected with the pre-existing Sino-Korean Paraplatform so as to form a great platform to be termed Chinese Protoplatform. However, this needs to be further verified especially by the work to be done in northern Qinling, northern Qilian and other regions.

According to the data available, the northern boundary of the Chinese Protoplatform was approximately situated along the Inner-Mongolian Axis, along the northern margin of northern Tianshan; as for the southwestern boundary, it was represented by the Red River Deep Fracture in the southern segment and probably by the Jinsha River Deep Fracture in the central segment. Further west the boundary coincided with the southern margin of the Kunlun Mts., while its southeastern boundary extended approximately from eastern Zhejiang via central Jiangxi, central Hunan and northern Guangxi to the area around Gejiu in southeastern Yunnan. Thus outlined, the Chinese Protoplatform was bordered on the north by the Central Asian-Mongolian Geosynclinal Region, on the southeast by the South China Geosynclinal Region, and on the southwest by the Paleo-Yunnan-Tibet Geosynclinal Region. In all these regions, the Sinian and Cambrian systems are usually characterized by continuously developed geosynclinal sediments. For instance, the Gongyanghe Group in western Yunnan consists of miogeosynclinal flysch and flysch-like sediments; the Sinian and Cambrian sequences within the South China Geosynclinal System are also mainly of miogeosynclinal flysch and flysch-like sediments, whereas in the Central Asian-Monogolian geosynclines there are both typical ophiolite suites and clastic and carbonate rocks of miogeosyncline type.

VI.3 THE ESTABLISHMENT OF THE XINGKAI CYCLE AND THE DISINTEGRATION OF THE CHINESE PROTOPLATFORM

VI.3.1 The Establishment of the Xingkai Cycle

The metamorphic complex constituting the Jiamusi Uplift in Northeast China is still assigned to early Proterozoic in the newly compiled *Geological Map of China* on a scale of 1:4,000,000 (Chinese Academy of Geological Sciences 1976) and the *Geological Map of Asia* on a scale of 1:5,000,000 (Chinese Academy of Geological Sciences 1975). However, as early as in 1974 we pointed out that they are not of the Early Proterozoic, but should belong to the Late Proterozoic, with their latest components even, most probably, involving the Lower Cambrian (Institute of Geology and Mineral Resources, unpublished data). The discovery of *Ediacara* fauna (Liu Xiaoliang 1978) from the Mashan Group in Heilongjiang Province has proved this inference. According to the data obtained by geologists of the Soviet Union (Okuneva and Rechena 1973) in the area near the Xingkai Lake in the southeastern part of the Jiamusi Uplift, the Lower Cambrian is distinctly unconformably overlain by the Middle Cambrian molasse containing trilobites. The Lower Cambrian is miogeosynclinal, rich in trilobites and archeocyathids, followed downwards by Sinian rocks. In that part of the Lesser Hinggan Mts. within the territory of the Soviet Union, all the miogeosynclinal sediments of the Sinian and Early Cambrian constitute approximately NS-trending, closely spaced linear folds, indicating that the Jiamusi Uplift (being a part of the Bureya Massif) was folded and consolidated at the end of Early Cambrian. As this fold is well developed in the area around the Xingkai Lake and represents the first Paleozoic orogenic phase, hence comes the name: the Xingkai Orogeny, equivalent to the Salair Orogeny, is known to have occurred in the Sayan Range between the Middle and Late Cambrian.

The Xingkai folding might also play quite an important role in the Yunnan-Tibet geosynclines. Based upon the facts that the Gongyanghe Group in western Yunnan is unconformably overlain by the fossiliferous Late Cambrian Baoshan Formation, and in conjunction with the geological data available from the Indo-China Peninsula, we believe that the huge Indosinian Massif and the folded basement of the Shan Highland are all products of the Xingkai Cycle. Qin Deyu holds the view that the basement of the Yunnan-Tibet geosynclines consists of Xingkaiides. Thus it is reasonable to suggest that the Qiangtang Plateau and Nam Lake Block possess a Xingkaiides basement (maybe partly Caledonian).

VI.3.2 The Disintegration of the Chinese Protoplatform

The folding and uplifting of geosynclines and the formation of new continental crust in some regions will in the long run be followed by taphrogeny and disintegration and by the formation of new geosynclinal belts in other regions. With the formation of the Xingkai fold belts in the Sayan Range, in the lesser Hinggan Mts., in western Yunnan, in Tibet, in Indo-China and elsewhere, the Chinese Protoplatform began to suffer rifting and disintegration. This is shown by the following facts:

a) In the Khariktau region of the western segment of the Chinese Tianshan Mts., the Sinian and Cambrian Systems are characterized by platformal sediments; and the Ordovician System, by miogeosynclinal sediments. It was not until the Silurian period that geosynclinal volcanism accompanied by jasper rocks was developed. In the Beishan Mt. region to the east of the eastern Tianshan Mts., the Sinian and Cambrian Systems are represented principally by platformal sediments, in which phosphorus-bearing formations analogous to those from the Yangtze Paraplatform were developed in the uppermost part of the Sinian, and in the lower and middle parts of the Cambrian. Not until the Ordovician and Silurian period did typical geosynclinal volcanic formations begin to develop.

b)　In the Qilian Geosynclinal System, the Lower Cambrian is usually absent. Beginning from the Middle Cambrian, eugeosynclinal sediments including volcanic formations were developed, while ophiolitic suites were formed in several zones.

c)　In the East Kunlun Geosyncline, the Sinian System consists of platformal sediments, the Cambrian being unknown. Real geosynclinal volcanic eruptives began to be deposited in the Ordovician.

d)　In the Qinling Geosynclinal System, platformal sediments dominate the Sinian. Since the Middle Cambrian, miogeosynclinal sediments began to develop, while the Silurian sequence is characterized by large amounts of volcanic rocks of great thickness.

The facts mentioned above demonstrate that all the geosynclines located in central and western China, including the Tianshan, Kunlun, Qilian, and Qinling Mts., were regenerated through taphrogeny of the Chinese Protoplatform after the Xingkai Cycle. The Sino-Korean Paraplatform, the Tarim Platform and the Yangtze Paraplatform, which persisted in the Paleozoic Era, are nothing but remnants of the disintegrated Chinese Protoplatform. Thus, it can be seen that the Xingkai Orogenic Cycle laid down the foundation for the development of the geotectonic framework of China since Paleozoic times.

VI.4　THE EARLY NEOGAIC MEGACYCLE (PALEOZOIC)
——THE FORMATION OF PAL-ASIA

After the Xingkai Cycle, the geotectonic pattern of China entered a new stage of development. To the north of the Tarim-Sino-Korean Paraplatform, the Central Asian-Mongolian Geosynclinal Region came into being, while the Kunlun-Qinling Geosynclinal Region occurred to the south. Southeast of the Yangtze Paraplatform was the South China Geosynclinal System, while southwest thereof was the Sanjiang Geosynclinal System. It is possible that northern Tibet, western Yunnan, Burma, Thailand and Cambodia used to belong to the Xingkaiides. The South China Sea Platform to the south of the South China Geosynclinal System might have also belonged to the Xingkaiides connecting northern Tibet, Burma and Cambodia.

The geotectonic evolution of China in the Paleozoic Era after the Xingkai Cycle is characterized by the continuous shrinkage of the geosynclinal mobile belts and the constant expansion of the stable platform regions. With the eventual closure of the Central Asian-Mongolian Geosynclinal Region by the Variscan Orogeny, the Siberian Platform was joined together with the Sino-Korean Paraplatform, forming the forerunner of the great Asian continent——Pal-Asia.

This consolidation process may be further divided into the Caledonian and Variscan Cycles. The Caledonian Cycle closed the South China Geosynclinal System, leading to the expansion of the platform in the southern part of China; essentially at the same time it closed the Qilian Geosynclinal System, giving rise to the expansion of the Tarim-Sino-Korean Paraplatform. Caledonian folding occurred in the Sanjiang Geosynclinal System on a much smaller scale, whereas to the north of the Sino-Korean Paraplatform and along the southern wing of the Inner Mongolian Geosynclinal Sysem, Caledonian folding took place in a narrow belt, forming an early geosynclinal orogenic cycle of the main Variscan Cycle.

The Variscan Cycle represents a major orogenic cycle in eastern and central Asia. The great Central Asian-Mongolian Geosynclinal Region, including the Altai, Junggar, the Tianshan and the Kunlun Mts., Inner Mongolia, Greater Hinggan Mts., as well as the region of Jiling and Heilongjiang Provinces, all turned into fold belts at the end of the Variscan Cycle (Permian), thus giving rise to the formation of the great continent, or Pal-Asia.

The formation of Pal-Asia gave the paleo-tectonics of China and its adjacent areas a completely new aspect. By this time, most of the Chinese terr tory and its adjacent areas, including the marginal Pacific geosynclinal belt encircling Japan, Taiwan Island and the Philippines, became stable and/or substable. Shallow-water and continental clastic sediments of Mesozoic age came into being in many marine depressions and nonmarine basins, such as the Ordos, the Junggar, the Tarim and the Sichuan Basins and many others.

In recent years, a number of Chinese geologists have tried to interpret the evolution of geosynclines in China by the theory of plate tectonics. In our *Tectonic Map of China* we draw a number of translithospneric fractures (plate sutures). It is necessary to notice that in the Paleozoic Era, most regions of China seemed to be tightly connected with one another (excluding the Himalayas and the extensive Central Asian-Mongolian Ocean), having undergone no large-scale horizontal displacement. The existence of older median massifs among the folded belts also suggests that extensive oceanic crust did not take part in the formation of the basement of these geosynclines. Though ophiolitic suites have been found in the northern Qilian Eugeosyncline, it has been proved to be connected with the Sino-Korean Paraplatform by a corridor transitional belt in the north; and in the south, it borders on the Qilian Median Uplift (Neo-Proterozoic), which extends to the east to form the Qinling Axis. Thus it appears quite possible that the Qilian Geosyncline is an intercontinental geosyncline without an extensive oceanic crust. In the foregoing chapter on Deep Fractures, we mark out such deep fracture belts as those located in northern Qilian, Jinsha River-Red River area, eastern Kunlun, Bangong Lake and Nujiang River; they are in no way equal to the modern Benioff Zone of the West Pacific, or to the plate suture occurring along the Yarlung Zangbo River. If the Benioff Zone of the West Pacific and the ophiolitic zone of the Yarlung Zangbo

River represent suture lines between plates of the greatest order, then the deep fractures located in northern Qilian, and in the Jinsha and Red Rivers can only represent sutures of a lower order.

VI.5 THE LATE NEOGAIC MEGACYCLE (MESOZOIC-CENO-ZOIC)——THE FORMATION AND DEVELOPMENT OF THE MARGINAL-PACIFIC TECTONIC DOMAIN AND THE TETHYS-HIMALAYAN TECTONIC DOMAIN

After the Late Triassic epoch-making Indo-Sinian Movement took place, the geotectonics of China again entered a new stage of development, the stage of intensive activities of the Marginal-Pacific and Tethys-Himalayan tectonic domains.

VI.5.1 The Development of the Marginal-Pacific Tectonic Domain

The formation of the Marginal-Pacific Tectonic Domain resulted from the drastic activity along the Benioff Zone in the West Pacific between the Pacific Ocean and the Asian continent. This tectonic domain comprises an inner belt and an outer belt. The inner belt is mainly a tectonic belt of Cenozoic age, including the West Pacific island arcs and marginal sea basins. The outer belt is essentially a tectonic belt of Mesozoic age, including the Mesozoic marginal Pacific geosynclinal fold belts and the Meso-Cenozoic activization belts of the continental margin in eastern Asia. The latter are superimposed on the older tectonic elements of different geological ages, containing, from north to south, the Inner Mongolian-Greater Hinggan Fold System, the Variscan Jilin-Heilongjiang Fold System, the Sino-Korean Paraplatform, the eastern segment of the Qinling Fold System, the Yangtze Paraplatform and the Caledonian South China Fold System, giving rise to strong folding and fracturing of the blanket sediments of platforms, to the formation of large amounts of volcanic rocks and granites, and to the occurrence of NE- and NEN-trending uplifts and downwarps.

The Marginal-Pacific Tectonic Domain itself has experienced three stages of polycyclic development, i.e. the Indosinian, the Yanshanian, and the Himalayan.

In the Paleozoic Era the Marginal-Pacific Geosynclinal Belt of eastern Asia was basically an Atlantic-type geosyncline, with the eastern margin of the continental crust of China representing a stable Atlantic-type margin. Consequently, the Marginal-Pacific Geosynclinal Belt by that time seemed to have no significant influence on the continent of China. But when entering into the Indosinian Cycle, and with the formation and development of the Benioff Zone in the West Pacific, the Marginal-Pacific geosynclines in eastern Asia had been converted into Pacific-type geosynclines, and the eastern margin of the continental crust of China had accordingly been converted into a Pacific-type continental margin undergoing an evolutional process ranging from the Andes type in the Mesozoic Era to the island arc-marginal sea type in the Cenozoic Era.

During the Indosinian tectonism, the strong compression along the Benioff Zone in the West Pacific between the Pacific Ocean and the Asian continent accounted for initiation of the intensive activity of the Marginal-Pacific Tectonic Domain in eastern China. It closed, on one hand, the Indosinian geosynclines in Japan and elsewhere, resulting in the formation of the first paired metamorphic belt——the Hida-Sangun paired metamorphic belt (Miyashiro, 1972)——and transformed the stable continental margin in eastern Asia into an active continental margin on the other hand, leading to the sea-floor spreading of the already consolidated Variscan folds, thus giving rise to the regeneration in Mesozoic times in the Sikhote-Alin region. Tectonic activization occurred in the eastern part of the Sino-Korean Paraplatform (Jiaoliao, Korea, and elsewhere) and in the eastern Yangtze Paraplatform (the lower reaches of the Changjiang River, northern Jiangsu and the Huanghai Sea), as well as in the Post-Caledonian Paraplatform of South China, and led to the formation of a complicated system of blanket folds accompanied by large-scale granitoid intrusions. Thereafter, quasi-molasse-type coal-bearing formations were accumulated in intermontane basins, on the continental margin of eastern China, as represented by the Anyuan coal measures (T_3), the Mentougou coal measures (J_{1-2}), etc. This is the first sequence of formations on the marginal activization belts of eastern China.

The Yanshanian Cycle was the most active stage in the Marginal-Pacific Tectonic Domain of eastern China, when the activity of the Benioff Zone in the West Pacific was further strengthened. With the closure of Mesozoic geosynclines in Japan, the second paired metamorphic belt——the Ryoke-Sanbagawa metamorphic belt——took shape, whereas eastern China witnessed strong folding, fracturing and large-scale calc-alkaline magmatism, as well as the most important endogenic metallogeny of China.

The Yanshanian Cycle may be subdivided into the early, middle and late stages. The Early Yanshanian, i.e. the Early and Middle Jurassic, served as the major subsiding stage for the Yanshanian Marginal-Pacific Geosynclinal Belt in eastern Asia, during which large-scale submarine volcanic eruption took place and ultrabasic rocks were formed in the Sikhote-Alin Geosynclinal Belt, while in the belt on the continental margin in eastern China the accumulation of post-Indosinian coal-bearing formations continued, being accompanied in places by volcanic eruptions. The Middle Yanshanian, i.e. the late part of the Middle Jurassic and the early part of the Early Cretaceous, represents a major folding stage for the Yanshanian Marginal-Pacific geosynclines in eastern Asia and also an important stage of magmatism in eastern China, when large amounts of volcanics and granites came into being. The late part of this stage was another important coal-forming stage in eastern China, of which the Fuxing and Jixi coal measures are good examples. The Late Yanshanian, i.e. the late stage of Early Cretaceous to the early stage of Late Cretaceous, was the final vanishing stage for the Yanshanian geosynclines in the West Pacific region and also the beginning of a stage during which the activization belts of the continental margin in eastern China, originally under compression, were transformed into regions of extension. Then, large-scale downwarping and rifting took place, resulting in sedimentary basins as represented by the Songliao Basin in Northeast China, while somewhat later in North China and South China medium- and small-sized down-faulted basins controlled by tensional fractures began to form, among which the third sequence of geological formations on the continental margin in eastern China began to accumulate. These are pre dominantly petroliferous formations and red beds with contiental evaporites.

The Himalayan Cycle represents a stage of development of island arc-marginal sea. In the Mesozoic Era, i.e. in the Indosinian-Yanshanian Cycle, the active continental margin in eastern China was basically a continental margin similar to that of the Andes type, but was gradually converted into a continental margin of island-arc and marginal-sea type when entering into the Himalayan Cycle. This conversion process may roughly be divided into two stages. In the Early Himalayan (late Cretaceous to Palaeogene), most parts of the Late Yanshanian Fold Belt bordering the continental margin of China (eastern margin of Asia), though partly separated from the Asian continent to form an island arc belt, had not yet entirely broken up and remained at the preludial stage of island-arc development. Therefore, within the inner side of the arcuate fold mountain system on this continental margin and in North China and northern Jiangsu, faulted basins of continental rift-valley type formed (and in the Eogene, their configurations and salient features were similar to those of the basin and range province of the western United States), in which salt-bearing red beds and petroliferous formations largely of continental facies were continuously accumulated, accompanied by frequent eruption of basaltic flow. In the Late Himalayan (since the Neogene), the Meso-Cenozoic fold mountain system on the eastern margin of Asia had been wholly separated from the continent, giving shape eventually to the present-day continental margin of island arc-marginal sea type. The geomorphology of the continental margin and sea-floor in eastern China fell into a pattern in this very period of time, where the East China Sea represents a developing geosynclinal belt, while the South China Sea is a marginal-sea basin with an oceanic crust. This general tectonic pattern reflects a tectonic background of expansion and subsidence since the Cenozoic. Nevertheless, if analyzed specifically, there is great difference from one one part of the country to another. For instance, South China is now under the influence of a NW-SE-trending compressive stress.

VI.5.2 The Development of the Tethys-Himalayan Tectonic Domain

The Tethys-Himalayan Geosynclinal Belt in southwestern China may obviously be divided into two parts by the Yarlung-Zangbo River Deep Fracture: the southern Tethys Geosynclinal Belt south of the Yarlung Zangbo River forms the northern marginal geosynclinal belt of Gondwanaland; the northern Tethys Geosynclinal Belt north of the Yarlung Zangbo River forms the southern marginal geosynclinal belt of the Asian continent. The Yarlung-Zangbo River Deep Fracture Zone in between the two geosynclines served as a suture zone between the two land masses after consumption of the oceanic crust of the Tethys Sea. This general tectonic history has also experienced a polycyclic developmental process of the Indosinian, Yanshanian, and Himalayan stages.

The Indosinian Movement here represents intensive compression between the oceanic or suboceanic crust of the Tethys Sea and the continental crust of Eurasia, during which major tectonism occurred east of both the Nujiang River Deep Fracture and the Jinsha River-Red River Deep Fracture, resulting in closure of the Qinling, Songpan-Garzê. and Sanjiang Geosynclines, and of those connected with them in Indo-China, thus giving rise to the Indosinian Geosynclinal Fold Region in southwestern Asia, perhaps the largest Indosinides in the world. At the back of this fold region, in the adjacent marginal regions of the original continent of Pal-Asia, large sedimentary basins predominantly of continental facies formed, such as Sichuan, Ordos, Qaidam, and Tarim sedimentary basins. The Sichuan and Ordos Basins were formely taken to be the result of Pacific tectonism, but, in fact, it would be more accurate to say that they are the result of the joint action of both the Tethys and the Pacific on the continent of Pal-Asia. Furthermore, in view of the great thickness of the Triassic and Jurassic deposits on the western margin of the Ordos Basin and in the piedmont depression belt of the Longmen Mt., it seems that the Tethys Tectonic Belt must have exerted much more influence on them.

The Yanshanian Cycle may also be divided into the early, middle, and late stages, of which the Late Yanshanian is the most important one. The Early Yanshanian Movement resulted in the formation of the Karakorum-Tanggula Fold System, while the Middle Yanshanian Movement resulted in the formation of the Nagqu Fold System, with the late Yanshanian Cycle being the last closing stage for the northern Tethys Geosynclinal Belt. The Late Yanshanian Movement, occurring in the late part of Late Cretaceous, reflected the last intensive compression between the oceanic crust or suboceanic crust of the Tethys Sea and the Asian continent, terminating the development of the northern Tethys Geosynclinal Belt, and forming the Late Yanshanian Gangdisê-Nyainqentanglha Fold System to the north of the Yarlung-Zangbo River Deep Fracture Belt. The Yanshanian Movement at the same time also induced first folding of the southern Tethys-Himalayan Geosynclinal Belt.

During the Himalayan Cycle, the oceanic crust of the Tethys Sea vanished, resulting eventually in a strong collision between the Indian subcontinent and the Asian continent along the suture line represented by the Yarlung Zangbo Valley, and thus the southern Tethys-Himalayan Geosynclinal Belt was closed and the Indian subcontinent merged with the Asian continent. This was an epoch-making and fundamental change in the tectonic aspect of western China. Since then, the principal expression of "movement of contradiction" was no longer seen between the Tethys Sea and the Asian continent, but transferred to the position between the Indian plate and the Asian plate.

The ever-expanding Indian Ocean pushed the rigid Indian Shield towards the mechanically weak Tethys-Himalayan Geosynclinal Fold Belt, producing a powerful compression which was concentrated in this fold belt and adjacent regions, thus strengthening and intensifying the movement of matter in the Earth's crust and upper mantle. The result was the rapid thickening of the crust and the large-scale uplifting of the Earth's surface, resulting in the formation of the great Qinghai-Tibet Plateau. This same compression was transmitted northwards along the southern margin of Pal-Asia, where rigid continental massifs (Tarim, Sino-Korean and Yangtze) resisted, producing a strong reaction, whereas the Paleozoic fold belts between these massifs, such as the Kunlun Mts., Qilian Mts., and Tianshan Mts. were "revived", resulting in strong block faulting and uplifting. In this way, several systems of "revived" Cenozoic mountain systems came into being. The fact that the late Cenozoic piedmont deposits in front of the Kunlun Mts., Qilian Mts., northern and southern Tianshan Mts. are quite similar to the Siwalik molasse of the Himalayas and are of the same age testifies to the conclusion that they were formed in one and the same stress field (Huang Jiqing 1979). In the meantime, the strong folding, overthrusting and imbricate structures found in the Tertiary and Quaternary sediments in these piedmont belts indicate that this region has been in a tectonic background of predominant compression since the Cenozoic, which is entirely different from that of predominant extension and subsidence in the Marginal Pacific Tectonic Domain in eastern China.

The uplift of the Qinghai-Tibet Plateau type and the formation of the revived mountain systems of the Tianshan type caused by a collision between the two continents and the powerful compression arising thereafter, as well as by the development of both the piedmont and the intermontane depressions which followed immediately, tend to serve as a special type of activization belt on the continental margin. In form, it is situated within the continent, but in fact, it is still located on the continental margin——the original marginal belt of the paleocontinent or neighbouring marginal belts. If the activization belt of the marginal Pacific continental margin in eastern China is named the Pacific type, then, correspondingly, the tectonically revived phenomena expressed in the Tethys-Himalayan Tectonic Domain in western China may be named the collision type, or the continental marginal activization belt of the Tethys-Himalayan type. From these facts, it can be seen that the Tethys-Himalayan continental marginal activization belt in western China involves not only the whole Qinghai-Tibet Plateau, but also exerts some influence on the Tianshan Mts. region.

The Mesozoic and Cenozoic Eras form an important period for the formation of oil and gas in China. Here we would like to make bold suggestions as to the rules of distribution of oil fields in China. The petroleum resources of China in the east, especially in the continental shelf, are of great importance, but it does not necessarily mean that western China should be ignored. The oil-bearing potentiality of the Tarim Basin is probably unique in the mainland of China. A general rule seems to exist: the farther both to the east and to the west of the line stretching from Kunming to Yinchuan, the richer the petroleum resource tends to be. That is to say, North China is superior to Ordos, and the continental shelf is superior to North China; Qaidam is superior to Ordos, and Tarim is superior to Qaidam. This "rule" is in any event worth further study.

VI.6 THE HIMALAYAN MOVEMENT AND ITS SIGNIFICANCE IN THE TECTONIC DEVELOPMENT OF CHINA

In a paper entitled *An Outline of the Tectonic Characteristics of China* (Huang Jiqing et al. 1977), we stressed the Indosinian Movement and its epoch-making significance in the history of tectonic development of the world, as well as of China, and summarized some extremely important characteristics of the Yanshanian Cycle in the Meso-Cenozoic tectonic evolution in China. Here, we lay special emphasis on the discussion of the Himalayan Movement and its important role in the tectonic evolution of China.

VI.6.1 Meaning and Subdivision

The lower age limit of the Himalayan Subcycle (or the upper age limit of the Yanshanian Subcycle) is, in general, placed between the Tertiary and the Cretaceous periods. However, the geological data obtained in the past decade or more tend to show that the intensive tectonic movement of the last episode of the Yanshanian Subcycle in China and its adjacent areas did not occur between the Cretaceous and the Tertiary periods, but in the Campanian-Maestrichitian stage within the late Cretaceous, with the strata of the terminal Cretaceous being commonly in a conformity or in a disconformity with those of the Tertiary. For instance, the folding period for the major cycle in the Yanshanian Sikhote-Alin Geosynclinal Fold System was the Maestrichitian stage (Krasnyi 1966), while in the last subduction of the Tethys Sea, the folding period for the major cycle in the Gangdisê-Nyainqentanglha Geosynclinal Fold System was the Campanian-Maestrichitian stage (Wang Naiwen, unpublished data). That is also the case for some Mesozoic and Cenozoic sedimentary basins of continental origin in China, as shown, for example, by an obvious unconformity between the Sifangtai Formation of the late part of the Late Cretaceous and the Nenjiang Formation of the early part of the Late Cretaceous in the Songliao Basin, with the former being in a conformity with the Cretaceous-Tertiary Mingshui Formation. In the Nanxiong Basin of Guangdong Province, there occurs an unconformity below the Late Cretaceous Nanxiong Formation which is in a conformity with the overlying Paleocene Shanghu Formation. That is to say, the boundary between the Himalayan and Yanshanian Subcycles would be rather inside the Late Cretaceous than between the Cretaceous and Tertiary Systems.

According to the data available, the Himalayan Subcycle is subdivided into two stages: the Early Himalayan Subcycle——the late stage of the Late Cretaceous to the middle (or initial) stage of the Miocene; and the Late Himalayan Subcycle, which ranges from the initial (or middle) stage of the Miocene to Recent. The Early Himalayan Subcycle comprises two important orogenic movements, the first one of which took place approximately from the Late Eocene to the initial stage of the Oligocene, being the major fold episode of the Himalayan Geosynclinal System; the second movement occurred in the initial (or middle) stage of the Miocene, giving rise to long-distance overthrusting, metamorphism and magmatic activities. This very movement represents a major folding phase of the Shimanto Geosyncline (Takachiho Movement) in Japan (Minato et al. 1965). The most important tectonic movement of the Late Himalayan Subcycle occurred in Pliocene-Pleistocene times, which is also a major folding phase for the Taiwan Fold System (Taiwan Movement, Ho 1975), and the main stage of uplifting for the Qinghai-Tibet Plateau also belongs to this movement. Consequently, the Himalayan Subcycle may be briefly summarized as follows:

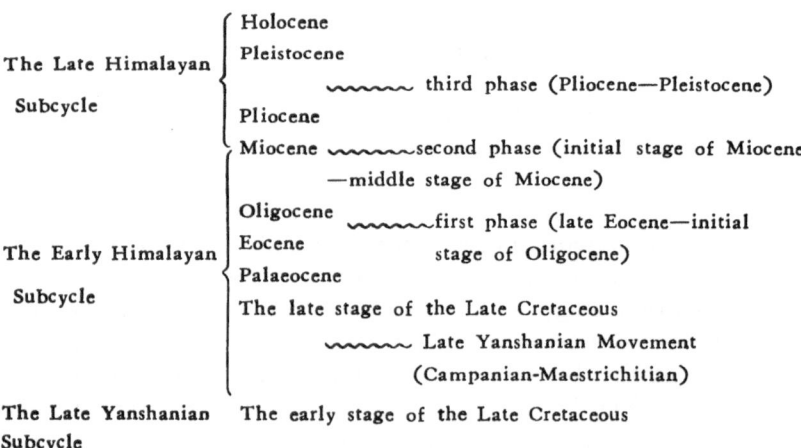

VI.6.2 The Tectonic Characteristics of the Himalayan Subcycle in Western China

The fact that the Himalayan Movement is extremely important both in the Himalayas and in Northwest China has long been universally accepted. Research work conducted in recent years has proved that the Himalayan Movement occurring in western Sichuan and western Yunnan is also of great importance; in particular, the final stage of the formation of nappe structures in the Longmen Mt. is considered as belonging to this Movement. The strong folding found in the red beds in western Yunnan was caused by the same movement, and in Lijiang, Heqing and elsewhere, Paleozoic and Triassic rocks are seen to rest on Tertiary red beds sometimes forming klippen.

The tectonic characteristics of the Himalayan Subcycle in western China are briefly summarized as follows:

1. The Himalayan Subcycle in western China is mainly characterized by intensive compression, resulting in the uplifting of old mountain systems, block mountains, as well as the magnificent Qinghai-Tibet Plateau.

2. The Himalayan Subcycle in western China consists of three major tectonic movements: that from Late Eocene to Early Oligocene, that from the end of Oligocene to the Middle Miocene, and that from the end of Pliocene to Early Pleistocene. For the early Himalayan Subcycle the most important tectonic movement was that occurring in late Eocene to early Oligocene, giving rise to a collision between Eurasia and the Indian subcontinent, and finally closing the Himalayan Geosyncline, whereas the Meso-Cenozoic red beds in northern Tibet and western Yunnan were folded intensively.

3. The most important late Himalayan Movement took place during the Plio-Pleistocene. This is the principal stage of uplifting for the Qinghai-Tibet Plateau, with the formation of huge piedmont depressions on both the southern and northern sides of the Plateau as well as to the north and south of the Tianshan Mts. Very strong folding and overthrusting occurred in the piedmont belts of these mountain ranges including the overthrusting of the Siwalik Hills (the Main Boundary Fault).

VI.6.3 The Tectonic Characteristics of the Himalayan Subcycle in Eastern China

The Himalayan Subcycle in eastern China may also be subdivided into early and late stages consisting of three phases. The first occurred between the Middle and Late Eocene, when the direction of the Pacific Plate motion changed from NWN to WNW (Uyeda and Miyashiro 1974). As is described above, that was also a stage for the collision between the Indian subcontinent and the Asian continent, as well as for the final closure of the Himalayan Geosyncline. The second one took place in the Early Miocene as manifested clearly in Taiwan Province and Japan. In the Taiwan Central Range, the Early Miocene and the Early Eogene sequences were simultaneously folded and metamorphosed (Ho 1975). In the East China Sea Basin, it appears that the Central Uplift represents also a fold belt of the Early Miocene which is linked with the Taiwan Central Range and the Shimanto Belt in Japan. On the mainland of China, the widespread occurrence of unconformities between the Oligocene and the Miocene sequences in a number of basins, such as the unconformity between the Guantao Formation and the

Dongying Formation in North China, and that between the Yancheng Formation and Sanduo Formation in northern Jiangsu, must also be taken as the result of this movement. Since the sediments concerned are predominantly of continental origin and are devoid of marine fossils, it is difficult to make a precise stratigraphical subdivision, and the age of the unconformity is vaguely designated as Oligocene-Miocene. The third movement occurred in the Pliocene and Early Pleistocene, as expressed distinctly in the Japanese and Taiwan island arcs, where the geosynclinal Cenozoic formations were folded to form the youngest fold belt in the West Pacific.

The tectonic characteristics of the Himalayan Movements in the Marginal Pacific Tectonic Domain in eastern China are quite different from those in the Tethys-Himalayan Tectonic Domain in western China. If the Himalayan Movement in western China is considered as being mainly characterized by strong compression, folding and uplifting, then in eastern China, it should be taken as being characterized by the formation of continental marginal fold belt and island-arc fold belt and by the development of back-arc tension basins by rifting. This is the well-known island-arc marginal sea type of tectonic configuration.

VI.6.4   The Geotectonic Evolution of China and Its Neighbouring Regions Since the Himalayan Subcycle

The general tectonic aspect since the Himalayan Subcycle, as expressed in the recent landforms of China, is typified in the thinning out of the Earth's crust with low and flat topography in the east, and in the thickening of the Earth's crust with high and rugged topography in the west. The maximum thickness of the Earth's crust may reach more than 70 km in the Qinghai-Tibet Plateau, and the Bouguer anomaly value there amounts to -550 milligals, with an elevation usually of 4000 m or more, forming the highest plateau in the world known as the Roof of the World. However, by reviewing the geological history, a better understanding of the important role of the Himalayan Movement will be acquired.

Since the Mesozoic, especially since the Himalayan Subcycle, with the gradual closure of the Tethys geosynclines, southwestern China had been transformed step by step from sea to land, and the Earth's crust had been becoming thicker and thicker, with the process of such a change reaching its culmination in the Lat Himalayan Subcycle. The geological data available indicate that in the Late Eocene, dating back to 40 my, the Himalayas and the western Tarim Basin were seated basically below the sea level. After the Eocene epoch, these regions had uplifted into land, their elevation was still near sea level (Huang Jiqing et al. 1980). It was not until the terminal Miocene that the average altitude reached several hundred meters above sea level. That is to say, by that time the thickness of the Earth's crust in the Qinghai-Tibet region was still quite small. Large amounts of geological data have proved that the modern geomorphology and thickness of the Earth's crust within the Qinghai-Tibet Plateau are the products of the latest Himalayan movements——the late stage of the Neogene (Huang Jiqing et al. 1980).

Contrary to the situation in western China, eastern China, with the formation of the West Pacific island arcs and marginal seas, had experienced a sort of process in which land had been changed into sea, and the thickness of the Earth's crust had been becoming smaller and smaller since the Late Mesozoic, particularly since the Cenozoic. Before the island arc-marginal sea took shape, Japan, Taiwan, and the Philippine Islands were directly connected with the mainland. In the Jurassic-Early Cretaceous times, that part of China to the east of the Greater Hinggan Mts.-Taihang Mts.-Wuling Mt. belt had generally been in a tectonic background predominantly of uplifting, with only medium- to small-sized intermontane down faulted basins being formed; to the west, Sichuan and Ordos had been in a tectonic background chiefly of downwarping, forming the huge Ordos, Sichuan, and central Yunnan basins. At the beginning of the late part of Early Cretaceous, especially since the Late Cretaceous (dated approximately at 120 my), the eastern part of China began to be faulted and downwarped, forming a series of faulted basins (Huang Jiqing et al. 1977), and it was not until the Neogene and Quaternary that the South China Sea, the East China Sea, the Huanghai Sea and the Bohai Sea began to take shape. In other words, during the Mesozoic Era there existed in eastern China thicker Earth's crust and corresponding higher topography. Since the Cenozoic, especially since the Neogene and Quaternary, great changes both in the ground surface and deep seated structures of eastern China have taken place due to the development of island arcs and marginal seas in the West Pacific, resulting in the modern pattern of distribution of lands and seas of China. Therefore, we may say that the difference in tectonic aspects between the eastern and western parts of China and the formation of the recent pattern of landforms resulted not only from the movement of the material in the sedimentary blanket of the Earth, but essentially from the movement of the material in the deeper part of the Earth's crust, particularly from the movement of the mantle material. These statements are supported by the published geophysical data. According to Dong Daliang (Deng Daliang, unpublished data) the velocity of the S-wave in the Earth's crust and upper mantle beneath the Qinghai-Tibet Plateau is quite small, illustrating that the top part of the mantle and the lower crust are in a partially melting state (see Fig. 30), which may be confirmed by the Quaternary volcanic activities and widespread distribution of hot springs in the Qinghai-Tibet Plateau (Geothermal Research Group Beijing University, unpublished data). It shows that since the Cenozoic Era, in the Qinghai-Tibet Plateau, strong differentiation and thermal expansion of the material from the Earth's curst and mantle has existed resulting in continuous upheaval of the terrain and continuous subsidence of the Moho surface. In eastern China, however, the up-arching of the Moho surface and the higher position of the layers of low seismic velocity tend to show that with the development of island arcs and marginal seas in the West Pacific, there has been a surging-up of mantle material since the Cenozoic, leading to the up-arching of the Moho surface as well as of the top layer of the asthenosphere.

What must be discussed in conclusion is in what direction (southwards or northwards) the Asian continent has in general moved since the Himalayan Subcycle. There are at present two completely opposite opinions concerning this problem. In view of the known facts, we are inclined to think that during this stage, the Asian continent moved from south to north.

As described in Chapter VI, in western China the sinistral Altun Fracture constitutes the northwestern boundary of the Qinghai-Tibet Plateau, and the dextral Yulong-Longmen Fracture (predominantly of compression with small dextral amplitude) makes up the southeastern boundary of the Qinghai-Tibet Plateau. In view of the occurrence of nappes in the piedmont of the Himalayas and the strong thrusting and overturning in the piedmont of the Qilian Mts. This general tectonic pattern has clearly demonstrated both the northward pushing and compression of the Indian Platform and the northward displacement of the Qinghai-Tibet Plateau as a whole, which have furthermore been strongly supported by the latest results of paleomagnetic studies. According to the report by Qian Fang et al. (1979), the present-day latitude of Yakou in the Kunlun Mts. is 35°42′N, but the paleolatitude of the Early Pleistocene Qiangtang Formation in N. Tibet was 32°34′N; the present-day latitude of the Jidagou Basin in the Yangbajin area in S. Tibet is 30°7′N, but the paleolatitude of the Pliocene and Early Pleistocene Jidagou Formation in that basin was 29°4′N.

In eastern China, the occurrence and development of a series of dextral tension-shear down faulted basins in North China and elsewhere since the Cenozoic, and the occurrence of a series of dextral shear fractures known from analysis on the mechanism of modern seismic foci all proved that the Asian continent has moved from south to north relative to the Pacific. In this regard paleomagnetic data have provided vigorous support. Liu Chun et al. (1978) pointed out, ''According to the paleomagnetic data obtained from the Pleistocene strata in eastern Asia their paleolatitude in the Middle and Early Pleistocene should fall within the range from 20° to 41°22′N. In other words, some regions of eastern Asia should belong to the zone of lower middle latitude. However, as for some individual localities, their paleolatitudes must be several degrees southward as compared with what they are in the present-day''. For instance, the modern latitude of the Kanto mountainous terrain of Japan is 36°N, but its paleolatitude is 28°12′; the modern latitude of the localities for paleomagnetic sampling in the northern part of Hakone and Izu is 35°N, and it paleolatitude is 31°42′N; the modern latitude for Hwatay, Panlyong and Mankyenli in northern Hamgyong of Korea is 41°N, but their paleolatitudes are 36°33′N, 35°N and 35°21′N, respectively; the modern latitude of Chifeng County in Liaoning Province is of 42°18′, and its paleolatitude, 36°06′N; the modern latitude of Lantian County in Shaanxi Province is 34°12′N, and its paleolatitude, 30°36′N; the modern latitude of the sampling locality on Cheleken Peninsula of the Soviet Union is 39°30′N, and its paleolatitude is 34°32′N.

"Due to the discrepancy caused by the current paleomagnetic methods, especially by the directional sampling in the field", the above-mentioned paleomagnetic data are by far not capable of examining quantitatively the rate of northward movement of the Asian continent, but the paleomagnetic studies made by different workers at different times and localities have unexceptionally shown that the Asian continent as a whole has been moving to the north at least since the Late Cenozoic, which could not but arouse special concern and attention.

In summary, the Himalayan Movement and its significance in the geotectonics of China may be briefly generalized as follows:

The Himalayan Subcycle began at the late stage of the Late Cretaceous but not at the initial stage of the Tertiary period, and may be subdivided into the early and late stages with three phases of important tectonic movements.

The Himalayan Movement occurring within the Marginal Pacific Tectonic Domain in eastern China and within the Tethys-Himalayan Tectonic Domain in western China shows different types and different characteristics.

The formation of both the latest tectonic framework and modern geomorphic features in China resulted from the tectonism of the Himalayan Subcycle, which represents not only the reflection of the movements of material of blanket sediments of the Earth but, in particular, the reflection of the movements of the material in both the deeper part of the Earth's crust and upper mantle. It originated from the joint action of the Indian Ocean Plate and the Pacific Plate exerted on the Asian continent, as well as from the reaction of the latter. Some geologists lay emphasis only on the role played by the Indian Ocean Plate. Such a viewpoint obviously needs further consideration.

Some basic views on the geotectonics of China have been given in this book. The position of China in global tectonics has determined the complexity and diversity of geological structures in China and the important role of China in the study of global tectonics as a whole. Though we have now acquired some preliminary knowledge of the rules governing the development of the geotectonics of China, it must be verified by future investigation, and there are still many aspects to be further explored.

It is to be noted that some previously published monographs on the geotectonics of China include a chapter on *Geotectonics and the Rules Governing the Distribution of Mineral Resources*. Considering that the discipline of metallogeny, though closely connected with geotectonics, is a special subject, it would be more appropriate to leave economic geologists handle this. We shall, however, certainly be happy if the fundamental knowledge given here is of some help to economic geologists.

The ring-shaped structures and the very striking lineaments obtained from the study of Landsat photographs are all shown on the *Tectonic Map of China* on a scale of 1:4,000,000. The mechanism of their formation awaits, of course, further thorough investigation.

References

Antonjuk R. M. et al. (1977) Struktury i evoljutsija zemnoj kory Tsentral'nogo Kazakhstana. Geotektonika, 5 pp. 71—82

Aubouin J. (1965) Geosynclines. Development in geotectonics, Vol. I, Elsevier, New York

Bakhtejev M. K., Filatova N. I. (1969) Ob osobennostjakh zakljuchitel'nogo etapa geosinklinal'nogo pazvitija Dzhungaro-Balkhashskoj geosinklinal'noj sistemy. Geotektonika, 3 pp. 75—87

Beljajevskij N. A. (1974) Zemnaja kora v predelakh territorii SSSR. Moskva. Nauka, p. 280

Belousov V. V. (1964) Vavlenie tektonicheskoi aktivizatsia v razvitii zemnoi koryi, aktivizirovannyi zonyi zemnoi koryi, Nauka, pp. 7—13

Bernoulli D., Laubscher H. P., Trumpy R., Wenk E. (1974) Gentral Alps and Jura Mountains. In: Mesozoic-cenozoic Orogenic Belts (Collat. and Edit. Spencer A. M.). Scottish Academic Press, Edinburgh London, pp. 85—108

Bogolepov K. V. (1973) O nekotorykh problemakh orogeneza. Trudy In-ta geol. i geofiz. SO AN SSSR, vyp. 173, pp. 8—17

Burchfield B. C., Davis G. A. (1972) Structural framework and evolution of the southern part of the Cordilleran orogen, western United States. Am. Jour. Sci., 272, pp. 97—118

Burk C. A., Drake C. L. (eds.) (1974) The geology of continental margins. Springer, Berlin, Heidelberg, New York

Burtman V. S. (1975) Structure geology of Variscian Tienshan, USSR. Am. J. Sci., 275A, pp. 157—186

Chang Chengfa, Zheng Shilang (1973) Tectonic features of the Mount Jolmo Lungma Region in Southern Tibet, China. Sci. Geol. Sin., 1 (in Chinese), pp. 1—2

Chao Kuokuang (1965) Preliminary observation on the Cenozoic stratigraphy and structure of Li-Chiang and Ta-Li Regions Northwest Yunnan. Geol. Rev., 23(5) (in Chinese), pp. 345—359

Chao Tsungpu (1956) Petrochemical study of the Cenozoic basaltic rocks in Eastern China. Acta Geol. Sin., 36(3) (in Chinese), pp. 315—367

Chen Bingwei, Ai Changxing, 1978. Some problems concerning the geotectonics in South China. Research on the Geology and Mineral Deposits, No. 19, pp. 66—83

Chen Guoda (Chen Kuo-ta) (1956) Examples of "activizing region" in the Chinese platform with special reference to the "Cathaysia" problem. Acta Geol. Sin., 36(3), pp. 239—271

Chen Guoda (1959) Theory of Progression with transformation between active and "stable" regions of the Earth's crust. Acta Geol. Sin., 39(3) (in Chinese), pp. 279—292

Chen Guoda (1960) Actixization of flatforms and its significance to ore-prospecting. Geological Publishing House, Beijing, China (in Chinese)

Chen Guoda, Chen Jiachao, Wei Bolin, Xue Jiamao, Liu Yishuan, Wen Shenji, Wei Zhaoling, Hu Hoyan (1975) A brief review on the geotectonics of China. Sci. Geol. Sin., 3 (in Chinese), pp. 205—219

Chen Haoshou (1975) On the isotopic ages of some granites and metamorphic rocks from

Northwest China. Acta Geol. Sin., 1 (in Chinese), pp. 45—60

Chen Ziqiang (1979) A discussion of the Dabie Movement and its duration of activity. 2nd All China Congr. Tecton. (unpublished)

Cheng Yuchi, Chung Futao, Su Yungjun (1973) The Pre-Sinian of Northern and Northeastern China. Acta Geol. Sin., 1 (in Chinese), pp. 72—81

Chinese Academy of Geological Sciences (1975) Geological map of Asia, on a scale of 1: 5,000,000. Cartogr. Publ. House, China

Chinese Academy of Geological Sciences (1976) Geological map of the People's Republic of China, on a scale of 1:4,000,000. Cartogr. Publ. House, China

Chung Futao (1975) K-Ar isochron ages of Precambrian rocks in Northern and Northeastern China. Geochemica, 2 (in Chinese), pp. 114—122

Compilation Group of the Regional Stratigraphic Table of Heilongjiang Province (1979) Regional stratigraphic table of Northeast China, 5 Fascicule Heilongjiang Province, p. 300. Geological Publishing House, Beijing, China

Compilation Group, Xinjiang Geological Bureau (1978) Geological and tectonic characteristics of Tianshan Mt. in China. A Collection of Papers for International Exchange, Section No. 1 (Regional Tectonics and Geomechanics), Geological Publishing House, Beijing, China. pp. 188—197

Cong Bolin, Ye Danian, Zheng Xuezheng, Jin Chengwei (1974) The mean atomic weights of rocks and geological problems. Sci. Geol. Sni., 1 (in Chinese), pp. 43—58.

Cong Bolin, Zhang Zhenhua, Zheng Xuezheng (1977) Petrochemistry of Mesozoic Volcanic Rocks in Eastern China and Its Geological Significance. Sci. Geol. Sin., 3 (in Chinese), pp. 245—259

Dai Wensai (1977) Evolution of Celestial Bodies. Science Press, Beijing, China. (in Chinese)

Deng Chitung, Wang Kelu, Wang Yipeng, Tang Hanjun, Wu Yuwen, Ding Menglin (1973) On the tendency of seismicity and their geological set-up of the seismic belt of Shanxi graben. Sci. Geol. Sin., 1 (in Chinese)

Deng Wanming (1978) A preliminary study on the petrology and petrochemistry of the Quaternary volcanic rocks of Northern Tibet Autonomous Region. Acta Geol. Sin., 52(2) (in Chinese), pp. 148—162

Ding Guoyu, Li Yongshan (1979) Seismicity and the recent facturing pattern of the Earth's crust in China. Acta Geol. Sin., 53(1) (in Chinese), pp. 22—34

Drake C. L., Ewing M., Sutton G. H. (1959) Continental marg ns and geosynclines: the east coast of North America of Cape Hatteras. Phy. Chem. Earth, 3, pp. 110—98

Drake C. L., Nafe J. E. (1968) The transition from ocean to continent from seismic refraction data, In: the Crust and Upper Mantle of the Pacific Area (Eds. Knopoff L., Orake G. L., Hart P. J.). Geophysical Monograph 12, American geophysical Union, Washington, D. C.

Esenov, Sh. E., Shlychyi E. D. (1972) Geologija SSSR. t. 20, Tsentral'nyj Kazakhstan. Nedra, M. p. 380

Esenov Sh. E. (1974) Geologija SSSR. t. 41, Vostochnyj Kazakhstan. Nedra, M. p. 395

Esenov Sh. E. (1971) Geologija SSSR. t. 40, Juzhnyj Kazakhstan. Geologicheskoe opisanije, Nedra, M. p. 404

Fang Zhongjing, Ti Fengjie, Xiang Hongfa, Ding Menglin (1976) The characteristics of Quaternary movement along the middle segment of the old Tancheng-Luijang Fracture Zone and their seismogeologic conditions. Sci. Geol. Sin., 4 (in Chinese), pp. 291—322

Fitch T. J. (1972) Plate convergence, transcurrent faults and internal deformation adjacent to southeastern Asia and the western Pacific. J. Geophys. Res., 77, pp. 4432—4460

Gansser A. (1964) Geology of the Himalayas. Interscience Wiley and Sons, New York

Gansser A. (1966) The Indian Ocean and the Himalayas. A geological interpretation. Ecl. Geol. Helv., 59(2), pp. 831—848

Gao Zhenjia (1978) Two problems concerning the geological Structures of the Tarim platform of Xingjiang. 2nd All China Congr. Tecton., Vol I. (in Chinese) (unpublished)

Gao Zhenjia, Peng Changwen, Li Yongan, Zhu Chengshun, Zhang Sengul (1979) The stratigraphic division and correlation of the Sinian and Cambrian in Kuruk Tagh, Xinjiang. Rep. Precambrian Geol. Xinjiang. (in Chinese) (unpublished)

Geologicheskij Institut Akademii Nauk SSSR, Ministerstvo geologii SSSR (1966) Tektonicheskaja karta Evrasii, 1:5,000, 000. Moskva

Geothermal Research Group, Institute of Geology, Academia Sinica (1979) Report on the data of terrestrial heat flow in North China Plain and adjacent regions and its study. Sci. Geol. Sin., 1 (in Chinese), pp. 13—21

Giese P., Prodehl C., Stein A. (1976) Explosion seismology in Central Europe. Springer, Berlin, Heidelberg, New York

Grachev A. F. (1977) Riftovye zony zemli. Nedra, p. 248

Grass L. G. (1976) Origin and emplacement of ophiolites, geodynamic today—a review of the Earth's dynamic processes, The Royal Society, London. pp. 55—65

Group of K-Ar Geochronology, Institute of Geology, Academia Sinica (1979) K-Ar dating and division of the Himalayan Movement in Southern Xizang. Sci. Geol. Sin., 1 (in Chinese), pp. 13—21

Guangzhou Seismological Party, State Seismological Bureau, under the direction of Chen Guoda, (1977), Tectonic map of China (1:4,000,000) (in Chinese), Cartographic Publishing House, Beijing, China

Guo Lingzhi, Shi Yangshen, Ma Ruishi (1980) The geotectonic framework and crustal evolution of south China. Scientific Papers on Geology for International Exchange, Prepered for the 26th International Geological Congress, No 1, Tectonic Geology and Geological Mechanics, Geological Publishing House, Beijing, China. (in Chinese) pp. 109—116

Guo Yong ling (1963) A preliminary recommendation about the tectonic regionization in Area from Guaxian to Zoige in Western Sichuan, Geol. Rev., 21(1) (in Chinese), pp. 6—11

Helde T. W. C., Uyeda S., Kroenke L. (1977) Evolution of the western Pacific and its margin. Tectonophysics, 38(1—2)

Ho C. S. (1975) An introduction to the geology of Taiwan, explanatory text of the geologic map of Taiwan. The Ministry of Economic Affairs, Taipei, Taiwan, p. 153

Hong Youchong, Yang Tzuchiang, Wang Shihtao, Wang Szu'en, Li Yu Kuei, Sun Mengrung, Sun Hsiangchun, Tu Naichui (1974) Stratigraphy and palaeontology of Fushun Coal Field, Liaoning Province. Acta Geol. Sin., 2 (in Chinese), pp. 113—158

Hsü Jen (1976) On the discovery of Glossopteris Flora in Southern Xizang and its significance in geology and palaeogeography. Sci. Geol. Sin., 4 (in Chinese), pp. 323—331

Hu Rao (1979) Ancient platemotion and evolution of the geosyncline in the Ondor Miao area of Inner Mongolia discussed in the light of the geological features in the area. 2nd All China Congr. Tecton., Vol I. (Abstr) (unpublished data)

Huang Jiqing (Huang T. K.) (1945) On major tectonic forms of China. Geol. Mem. Natl. Geol. Surv. China Ser. A (20), 212p.

Huang Jiqing (Huang T. K.) (1960) The main characteristics of the structure of China; preliminary conclusions. Sci Sin., 4, pp. 492—544

Huang Jiqing (Huang T. K.) (1978) An outline of the tectonic characteristics of China. Ecl. Geol. Helv., 71(3), pp. 611—635

Huang Jiqing (Huang T. K.) (1979) A preliminary discussion on the polycyclic development of Geosynclinal foldbelts. Sci. Sin., 4 (in Chinese), pp. 384—397

Huang Jiqing (Huang T. K.), Chang Chengkun, Chang Chihmeng, Chen Kuoming (1965)
On eugeosynclines and miogeosynclines of China and their polycyclic Development.
Professional Papers, Section C, Regional Geology and Structural Geology, No. 1, Chi-
na Industry Press, Beijing, 71p.

Huang Jiqing, Chen Bingwei (1980) On the formation of Pliocene-Quaternary molasses in
Tethys-Himalayan Tectonic Domain and its relation with the Indian plate motion.
Scientific Papers on Geology for International Exchange, Prepared for the 26th In-
ternational Geological Congress 1 (Tectonic Geology and Geological Mechanics),
Geological Publishing House, Beijing, China. pp. 1—14

Huang Jiqing (Huang T. K.), Hsu K. C. (1936—37) Mesozoic orogenic movements in
the Pinghsiang Coalfield, Kiangsu. Bulletin of the Geological Society of China, XVI,
pp. 177—194

Huang Jiqing (Huang T. K.), Jiang Chunfa (1962) Preliminary investigation on the evolution of the Earth's crust from the point of view of polycyclic tectonic movements.
Acta Geol. Sin., 42(2) (in Chinese), pp. 105—152

Huang Jiqing (Huang T. K.), Jén Chishun, Jiang Chunfa, Chang Chihmeng, Hsu Chihchin
(1977) An outline of the tectonic characteristics of China. Acta Geol. Sin., 2 (in
Chinese), pp. 117—135

Huang Jiqing (Huang Chi-ching), Jen Chishun, Jiang Chunfa, Chang Chihmeng, Chang
Chengkun (1974) Some new observations on the geotectonic characteristics of
China. Acta Geol. Sin., 1 (in Chinese), pp. 36—52

Institute of Geology, Academia Sinica (1959) An outline of geotectonics of China. Science
Press, Beijing, China (in Chinese)

Institute of Geophysics, Academia Sinica (1974) Reflected waves from the M. Discontinuity and the structure of Lin Fen Basin. Acta Geophys. Sin., 17(1) (in Chinese),
pp. 239—246

Institute of Geomechanics, Chinese Academy of Geological Sciences (1976) Tectonic
system map of the People's Republic of China. Cartographic Publishing House,
Beijing, China (in Chinese)

Jan M. Q., Kempe D. R. C. (1973) The petrology of the basic and intermediate rocks of
upper Swat, Pakistan. Geol. Mag., 11(3), pp. 285—300

Janov, E. N. (1977) Tipy podvizhnykh oblastej i stadii ikh razvitija. Cov. geol., 4, p.
8—19

Janshin A. A., Gnibidenko G. S., Kosyigin Y. A., Sergeev K. F., Soloviev, S. L., (1976) Stro-
jenije zemnoj kory i verkhnej mantii v zone perekhoda of Aziatskogo kontinenta k
Tikhomy Okeanu. Nauka. Sibirskoje otdelenije. p. 368

Jiang Chunfa, Zhang Qinggui, Zhang Yuxou, Zhu Zhizhin (1963) The presence of the
Indosinian Movement in the Eastern Qinling Geosyncline. Geol. Rev., 21(1) (in
Chinese), pp. 116—121

Jin Chengwei, Zhou Yushen (1978) Igneous rock belt in the Himalayas and the Gangdese
Arc and their genetic model. Sci. Geol.Sin., 4 (in Chinese), pp. 297—312

Juan V. C. (1975) Tectonic evolution of Taiwan, Tectophysics, 26, pp. 197—212

Khain V. E., (1974) Uchenie o Geosinklinaljakh na novom etape razvitija geologicheskoj
nauki. Vest. MCU. geol., 2, pp. 3—21

Khain V. E. (1977) Shest'desjat let sovetskoj geotektoniki. Geotektonika. 5, pp. 5—19

Khain V. E. (1973) Obshaja geotektonika, 1-e isdanije 1964. 2-e isdanije. pp. 511

Ke Yuan, Wu Zhen (1976) On the Upper Pre-Cambrian of Western Henan and its significance in stratigraphical correlation. Sci. Geol. Sin., 2 (in Chinese), pp. 151—168

Krasnyi L. I. (1966) Geologija SSSR. t. 19, Khavarovskij kraj i Amurskaja oblast'. Nedra,

M. p. 272

Laboratory of Isotope Geology, Guiyang (Kweiyang) Institute of Geochemistry, Academia Sinica (1977) On the Sinian geochrononological scale of China based on isotopic ages for Sinian strata in the Yenshan Region, North China. Sci. Sin., 20(2) (in English), pp. 151—161

Lee J. S. (1939) The geology of China. Murby, London, p. 528

Lee J. S. (1973) Crustal structure and crustal movement, Sci. Sin., 16(4), pp. 519—559

Le Pichon X., Framcheteam, J., Bonnin J. (1973) Plate tectonics. Developments in geotectonics, Vol VI, Elsevier, New York.

Li Chunyu (Lee Chunyu) (1975) Tectonic evolution of some mountain ranges in China, as tectonically interpreted on the concept of plate tectonics (in Chinese). Acta Geophys. Sin., 18(1), pp. 52—76

Li Chunyu (1978) History of the development of tectonics in Qinling and Qilian Mts. A Collection of Papers for International Exchange No 1 (Section on Regional Tectonics and Geomechanics), Geological Publishing House, Beijing, China (In Chinese)

Li Guopeng, Li Yusong (1973) The features of Mesozoic and Cenozoic tectonic developments in Eastern China and their relationship to seismicity. Sci. Geol. Sin., 3 (in Chinese), pp. 238—244

Li Jijun, Wen Shishuan, Zhang Qingsong, Wang Fubao, Zheng Benxing, Li Bingyuan (1979) A discussion on the period, amplitude and type of the uplift of the Qinghai-Xizang Plateau. Sci. Sin., 22(11) (in Chinese), pp. 608—616

Li Jiliang, Chen Changming, Gao Wenxue, Chen Ruijun, Huang Jiakuan, Tian Youxing, Liu Jiehan (1978) On the features of turbidite sequences in some regions of China. Sci. Geol. Sin., 1 (in Chinese), pp. 26—44

Li Pu, Liu Chun (1979) Paleomagnetic study of the Sinian system in from Eastern Gorge Districts of the Yangtze River. Acta Geophys. Sin., 22(3) (in Chinese), pp. 281—288

Li Siguang (Lee J. S.) (1973) Introduction to geomechanics. Science Press, China (in Chinese)

Li Yujao (Lee Yujao), Li Jie, Zhu Sen (1935) The Geology of the Ningchen Mt. Bulletin of the Institute of Geology, Academia Sinica, 11, pp. 1—379

Li Zhlyi, Cai Wenbo, Ding Menglin, Xu Haomin, Wang Yipeng (1974) A note on the seismo-geologic features in China. Sci. Geol. Sin., 4 (in Chinese), pp. 356—370

Liu Changan, Shan Jicai (1979) Basic characteristics of the ancient plate tectonics in the Mongolia-Okhotsk Sea (unpublished) (in Chinese)

Liu Chun, Liu Haishan (1965) Paleomagnetic study on some Sinian sequences in China. Sci. Geol. Sin., 1 (in Chinese), pp. 77—79

Liu Chun, Zhu Xiangyuan, Ye Sujuan (1977) A paleomagnetic study on the cave-deposits of Zhoukoudian (Choukoutian), the locality of Sinanthropus. Sci. Geol. Sin., 1 (in Chinese), pp. 26—33

Liu Chun, Zhu Xiangyuan, Ye Sujuan (1978) The geological significance of the paleomagnetic data from the pleistocene sequences of East Asia. (unpublished)

Liu Hongyun (Liu Hungyun), Sha Qingan, Hu Shiling (1973) The Sinian system in Southern China. Sci. Sin., 16(2) (in English), pp. 202—212

Liu Yuanlung, Wang Chienshen, Wu Chuanzhen, Zhou Wenhu (1977) A preliminary study of the crustal structure and the geological significance of the Central Portion of the Himalayan Mountain Range. Acta Geophys. Sin., 20(2) (in Chinese), pp. 143—149

Liu Xiaoliang (1978) The discovery of Metazoan fossils from the Mashan Group in Heilongjiang Province and Their geological significance. (unpublished) (in Chinese)

Lowell J. D., Genik G. J., Helson T. H., Tucker P. M. (1975) Petroleum and plate tectonics of the Southern Red Sea, In: (Eds. Fischer A. G., Judson S.) Petroleum and Global Tectonics, Princeton University Press, pp. 129—156

Luo Fazuo (1981) Talking about the cycle of Tianshan, In: Selected Papers from the Second All-China Congress on Tectonics,V. 1, Geological Publishing House, Beijing, China, pp. 18—27

Luo Zhili (1979) On the occurrence of Yangze old plate and its influence on the evolution of lithosphere in the southern part of China. Sci. Geol. Sin., 2 (in Chinese), pp. 127—138

Ma Xingyuan (Ma Hsingyuan) (1957) On the pre-Cambrian stratigraphy of the Sungshan Area, Henan, and the promblems of its correlation. Acta Geol. Sin., 37(1) (in Chinese), pp. 11—32

Ma Xingyuan (Ma Hsingyuan), Yu Chendong, Tan Yingchia, Yang Weizhan, Lee Dongshui, Wu Zhengwen (1961) Some fundamental problems in Chinese geotectonics.Acta. Geol. Sin., 41(1) (in Chinese), pp. 30—44

Magnetoelectric Sounding Group, the Seisinological Bureau of Lanzhou, National Seismological Bureau (1975) On the electric conductivity characteristics of the northern portion of the North-South earthquake belt of China. Acta Geophys. Sin., 19(1) (in Chinese), pp. 28—34

Manik Talwani, Pitman W. C. III (eds) (1977) Island arcs, deep sea trenches and back-arc basins. Am. Geophys. Union. Wash. 470p

Marinov N. A., Zonenshajn I. P., Blagonravov V. I. (1973) Geologija Mongol'skoj Narodnoj Respubliki. t. 1,2. Nedra, M. p. 583, p. 751

Minato M. et al. (1965) The geologic development of the Japanese Islands. Tsukiji Shokan Co., Japan. 442p

Minero-Petographic Laboratory Inst. Geol. Acad. Sin. (1977) A preliminary study of igneous activities in the region around Bohai Sea. Sci. Geol. Sin., 4 (in Chinese), pp. 322—342

Mirchink G. F. (1940) Osnovnyie zakonnomernosti razvitia zemnovo shara. Biull. MOIP., otd. geol., 3—4

Mitchell A. H., Reading H. G. (1969) Continental margins. geosynclines and ocean floor spreading. J. Geol., 77, pp. 629—646

Miyashiro A. (1972) Metamorphism and related magmatism in plate tectonics. Am. J. Sci. 272, pp. 629—656

Molnar P., Tapponnier, P. (1975) Cenozoic tectonics of Asia: Effects of a continental collision. Science, 189, pp. 419—426

Mossakovskij A. A. (1965) K voprosu ob orogennom etape pazvitija geosinklinal'nykh oblastej. Geotektonika, 2, pp. 3—6

Mueller S. (1974) The structure of the Earth's crust based on seismic data. Dev. Geotecton., 8, p. 391

Muratov M. V. (1977) Osnovnye tektonicheskije podrazdenija territorii Sovetskogo Sojuza. Geotektonika, 5, pp. 15—31

Muratov M. V. (1971) Osnovnye tipy geosinklinal'nykh progibov i ikh magmatizm. Isv. AN SSSR. Serija geol., 5, pp. 3—11

Okuneva O. G., Renina I. N. (1973) Biostratigrafia i fauna kembria primoria, Novosibirsk "Nauka", p. 38

Pan Yusheng (1981) Geotectonic configuration of Ahrin Area of Tibet. In: Selected papers from the Second All-China Congress on Tectonics, Vol. I, Geological Publishing House, Beijing, China. (in Chinese), pp. 10—17

Pavlov V. A., Sichev N. M. (1975) Gravity anomalies of the East Chinese Sea and Ryukyu Arc (Nansen). Geology and Geophisics, 1(181), pp. 146—152

Pejve A. V. (1956) Obshaja kharakteristika, klassifikatsija i prostranstvennoje raspolozhenije glubinnykh razlomov. Glavnejshije tipy glubinnykh razlomov. Stat'ja 1. Izvestija AN SSSR, Serija geologicheskaja, 1, pp. 90—105

Pejve A. V. (1956) Svjaz' osadkonakoplenija, skladchatosti, magmatizma i mineral'nykh mestorozhdenij s glubinnymi razlomami. Glavnejshije tipy glubinnykh razlomov. Stat' ja 1. Izvestija AN SSSR. Cerija geologicheskaja, 2, pp. 57—71

Pejve A. V., Shtrejs N. A., Knitger A. L. et al. (1971) Okeany i geosinklinal'nyj protsess. Dokl. AN SSSR, 196, (3), pp. 657—659

Prodehl C., Ansorge edel J. B., Emter D., Fuchs K., Mueller S., Peterschmitt (1976) Explosion-seismology research in the Central and Southern Rhine Graben In: A case History in "Explosion Seismology in Central Europe" (Ed. P. Giese et al.), pp. 313—328

Puzyrev N. N., Mandelbaum M. M., Krylov S. V., Mishenkin B. P., Petrik, Krupskava G. V. (1978) Deep structure of Baikal and other continental rift zones from seismic data. Tectonics, 45(1), pp. 15—22

Qian Tang, Ma Xinghua, Wu Xihao, Pu Qingyu (1979) A study on the paleomagnetism of the late cenozoic pluvial and lacustrine deposits in the Tanggula Range and Eastern Kunlun Areas. (unpublished data)

Qiao Xiufu, Geng Shufang (1981) On Late Pre-Cambrian plate tectonics of South China. In: Contributions to the Tectonics of China and Adjacent Regions (Edit. Huang Jiqing, Li Chunyu), Geological Publishing House, Beijing, China. (in Chinese) pp. 77—91

Qin Deyu, Li Guangcen (1976) Geological report on work conducted along the Gela Investigation Route (unpublished) (in Chinese)

Ren Jishun (Jen Chi-shun), Chu Chingchuan (1966) On the occurrence of geosynclinal indosinides in the Lanping-Weihsi Region, Western Yunnan. Acta Geol. Sin., 46(2) (in Chinese), pp. 182—200

Ren Jishun (Jen Chishun) (1964) Preliminary study on some Pre-Devonian geotectonic problems of Southeastern China. Acta Geol. Sin., 44(4) (in Chinese), pp. 418—431

Roberts D. G. (1974) Structural development of the British Isles, the continental margin, and the Rockall Plateau. In: The Geology of Continental Margins, (Edit. Burk C. A., Drake C. L.), Springer-Verlag, Berlin, Heidelberg, New York, pp. 343—356

Rotarash I. A., Grediushko I. A. (1974) Istorija formirovania stroenie serpintinitovovo melanzha zaisanskoi skladchatoi oblatti, Geotectonika, 4, pp. 73—79

Salun, S. A. (1977) Osnovnyje cherty tekto-niki i istorija razvitija Sikhote-Alinskoj skladchatoj sistemy. Geotektonika, 1, pp. 59—102

Schuchert C. (1923) Sites and natures of the North-American geosynclines. Bull. Geol. Soc. Am., 34, pp. 151—260

Sergejev K. F. (1976) Tektonika Kuril'skoj ostrovnoj sistemy. Nauka. p. 240

Seyfert C. K., Sirkin L. A. (1973) Earth history and plate tectonics. An introduction to historical geology. Harper and Row, New York

Section of Structural Geology, Institute of Geology, Chinese Academy of Geological Science (1979) Tectonic Map of China. Cartographic Publishing House, Beijing, China

Shanxi Regional Survey Party (1968) The Fengxian sheet and its explanatory notes (unpublished data)

Shao Xuezhong, Zhang Jiaru, Yang Xiaofeng, Zhang Xiaohua, Lei Shengli, Wang Qiming, Gao Wei'an (1978) An experimental study of the structure of the Earth's crust and upper mantle by converted waves. Acta Geophys. Sin., 21(2) (in Chinese), pp. 89—100

Shi Zheliang, Huan Wenlin, Wu Huangying, Cao Xinling (1973) On the intensive seismic activity in China and its relation to plate tectonics. Sci. Geol. Sin., 4 (in Chinese), pp. 281—293

Sichuan Provincial Geological Bureau (1974) Geological map (Qamdo sheet) on the scale 1:1,000,000 (unpublished)

Sinkha Roj S. (1977) Metamorfizm i tektonika Gimalajev na primere Vostochnykh Gimalajev. Geotektonika, 2, pp. 66—74

Sonder R. A. (1956) Mechanik der Erde, Elemente und Studien zur tektonischen Erdgeschichte. Stuttgart, pp. VII-29

Songshan Team, Wuhan Geologic College (1977) The Pre-Cambrian deformation history and paleo-structural types of the Songshan Area, China. Sci. Geol. Sin., 2 (in Chinese), pp. 105—117

Spencer A. M. (ed) (1974 Mesozoic-Cenozoic orogenic belts. Data for orogenic studies. Scottish Academic Press, Edingburgh London

Spencer E. W. (1977) Introduction to the structure of the Earth. McGraw-Hill, New York

Stille H. (1940) Einfuhrung in den Bau Nordamerikas. Borntrager, Berlin, p. 717

Stille H. (1924) Grundfoagen der vergleichenden Tektonik. Gebruder Borntraeger, Berlin, p. 443

Sugimura A., Uyeda S. (1973) Island arcs; Japan and its environs. Developments in geotectonics. Elsevier, New York

Tapponnier P., Molnar P. (1976) Slipline field theory and large-scale continental tectonics. Nature (London), 264(5584) pp. 314—324

Tectonic Map Compiling Group Inst. Geol. Acad. Sin. (1974) A preliminary note on the basic tectonic features and their developments in China. Sci. Geol. Sin., 1 (in Chinese), pp. 1—17

Tektonika Uralo-Mongol'skogo skladchatogo pojasa. (Trudy soveshchenija) 1974, Nauka. pp. 181

Teng Jiwen (Teng Chi-wen) (1974) Deep reflected waves and the structure of the Earth's crust of the eastern part of Qaidam Basin. Acta Geophys. Sin., 17(2) (in Chinese), pp. 122—135

Teng Jiwen (Teng Chi-wen), Feng Chinfen, Li Kinsun, Chen Hsuehpu, Wen Kunti, Chang Kiaju, Hsing Chengchun (1974a) Crustal structure of the central part of the North China Plain and the Hsintai earthquake. Acta Geophys. Sin., 17(2) (in Chinese), pp. 254—271

Teng Jiwen, Wang Guacheng, Liu Daohuang, Hsing Jiyun, Liang Wendou, Xu Shilin (1973) Caustal structure of the central part of the North China Plain and the Hsintai earthquake. Acta Geophys. Sin., 18(3), (in Chinese), pp. 196—207

Teng Jiwen, Yao Hung, Chou Hainan (1979) Crustal structure in the Beijing-Tienjin-Tangshan-Zhangjiakou Region. Acta Geophys. Sin., 22(3) (in Chinese), pp. 236—241

Terman M. J. (1973) Tectonic map of China and Mongolia. (1: 5,000,000). Geol. Soc. Am. Boulder, Colorado.

Tsvetkov A. J., Shvanov V. M. (1976) Ob odnom maloizvestnom tipe glaukofanovogo metamorfizma v Juzhnom Tjan'shane. Dokl. AN SSSR. 230(3)

Tujezov I. K. (1975) Litosfera Aziatsko-Tikhokeanskoj zony perekhoda. Nauka. Sibirskoe otdelenije. p. 232

Usov M. A. (1936) Fazyi itsikli tectogeneza zapadno-sibirskovo kraya, Tomsk, p. 209

Uyeda S., Miyashiro A. (1974) Plate tectonics and the Japanese Islands: a synthesis. Geol. Soc. Am. Bull., 85, pp. 1159—1170

Van Bemmelen R. W. (1977) The undation theory. Geologie en Mijnbouw, 56(3), pp. 263—269

Volkov V. V. (1966) Osnovchye zakonomernosti geologicheskogo razvitija gornogo Altaja Nauka. p. 162

Wang Chiayin (1974) On fracture zones. Sci. Geol. Sin., 2(in Chinese), pp. 146—160

Wang Henian (Wang Ho-nyen) (1961), On the discovery of flysch formation and discussion on the Pre-Cambrian Pan-chi Series, W. Hunan. Acta Geol. Sin., 41(1) (in Chinese), pp. 16—22

Wang Hongzhen (Wang H. C.) (1955) Geotectonic units of Eastern China, as specially viewed from the development of the Pre-Cambrian systems of the same region. Acta Geol. Sin., 35 (in Chinese), pp. 375—404

Wang Hongzhen (Wang H. C.) (1978) On th subdivision of the stratigraphic provinces of China. Acta Stratigr. Sin., 2(2) (in Chinese), pp. 81—104

Wang Kaiyuan and Shi Giaxin (1981) The main characteristics of the structural metamorphic zones of Western Yunnan. In: Selected Papers from the Second All-China Congress on Tectonics. Vol. 1, Geological Publishing House, Beijing, China. (in Chinese), pp. 79—89

Wang Quan, Liu Xueya (1976) Paleo-oceanic Crust of the Chilienshan Region, Western China and ts tectonic significance. Sci. Geol. Sin., 1(in Chinese), pp. 42—55

Wang Xiuzhang (1959) Mesozoic marine sequences in the Raohe Mesozoic Foldbelt, Northeast China. Sci. Geol. Sin., 2 (in Chinese), pp.50—51

Wang Y. L. (1963) Problems of the Sinian-Cambrian boundary of North China. Acta Geol. Sin., 43(2), pp. 116—135

Wang Zhi, Wen Guang (1957) Mednoe mestororjenie borfirovih rud Zhongtiaoshanskobo tipa. Acta Geol. Sin., 37(4) (in Chinese) pp. 401—413

Woll P. W., Fischer W. A. (eds) (1977) Proc 1st Annu. william T Pecora Memorial Symp. October 1975, Sioux Talls, SD US Government Printing Office, Washington DC

Wong W. H. (1926) Crustal movement in Eastern China. Proc. Pan. Pacif. Sci. Congress, Tokyo, I. pp. 467—475

Wu Haoruo, Wang Dong'an, Wang Liancheng (1977) The Cretaceous of Laze-Jiangze District, Southern Xizang. Sci. Geol. Sin., 3 (in Chinese), pp. 250—262.

Wu Changsheng (1979) The Xar Maron deep fracture. 2nd All China Congr. Tecton. (Abstr) (unpublished) (in Chinese)

Wu Jiyuan (1979) A discussion of the structural characteristics of the Lingshan Fold-Fracture Belt and its tectonic nature. 2 nd All China Congr. Tecton. (unpublished)

Wyllie P. J. (1971) The dynamic Earth: Textbook in geoscience, pp. 416

Xia Linqi (1976) The origin of spilite-keratophyre suite in Mian-Lue Regino and a preliminary note on the mechanism of differentiation at depth of its parent magma. Acta Geol. Sin., 1 (in Chinese), pp. 24—37

Xiao Xuchang, Chen Guoming, Zhu Zhizhi (1978) A preliminary study on the tectonics of ancient ophiolite in the Qilian Mountains, Northwest China. Acta Geol. Sin. 52(4) (in Chinese), pp. 281—295

Xu Jiawei (1978) A preliminary discussion on the horizontal displacement of Tancheng-Lujiang Fracture Zone and its geological and ore-prospecting significance. Res. Geol. Miner. Resour. 5, pp. 1—30

Yan Qinshang (1979) Marine inundation and related sedimentary environment of the Funing Group (Lower Paleogene) in the Jinhu Depression, North Jiangsu Plain. Acta Geol. Sin., 53(1), (in Chinese), pp. 74—84.

Ye Danian, Li Dazhou, Dong Gengfu, Qiu Xiuwen (1979) Metamorphosed 3T-phengire eclogite of group C in Xinyang Region of Henan Province, China. Sci. Bull. 5 (in Chinese), pp. 217—220

Ye Hong, Liang Yishan, Shen Liqi, Xiang Hongfa (1975) The analysis of the recent tectonic stress of the Himalaya Mountains Arc and its vicinities. Sci. Geol. Sin., 1 (in Chinese), pp. 32—48

Yen T. P. (1975) Lithostratigraphy and geologic structure. Geol. Palaeontol. SE Asia, 15, pp. 303—323

Yin Jixiang, Guo Shizeng (1976) On the discovery of the stratigraphy of Gondwana facies in northern slope of the Qomolangma Feng in Southern Xizang, China. Sci. Geol. Sin., 4 (in Chinese), pp. 291—322

Yin Jixiang (Yin Chi-hsiang), Kuo Shihtseng (1978) Stratigraphy of the Mount Jolmo Lungma and its north slope. Sci. Sin., 21(5) (in English), pp. 90—102

Yin Zhankun, Zhang Shouxin, Xie Cuihua (1965), Recommodation (draft) on the existing terminology of crustal movements in China. Geol. Rev. 23, Special Issue on the Terminology of Crustal Movements in China. (in Chinese), pp. 5—19

Yin Zhanxun, Zhang Shouxin, Xie Cuihua (1978) On folding episodes. Science Press, Beijing, China. (in Chinese)

Ying Szuhuai (1973) Magmatic, metamorphic and migmatic rocks of the Mount Jolmo Lungma in Southern Tibet, China. Sci. Geol. Sin., 2 (in Chinese), pp. 103—132

Yu Deyuan (1954) Geotectonics and distribution of mineral resources in China. Acta Geol. Sin., 37(3) (in Chinese), pp. 257—270

Zajchev N. S. (1974) Tektonika Mongol'skoj Narodnoj Respubliki. Nauka. p. 284

Zeng Rongsheng (Tseng Jungsheng) (1973) Gravity compensation of the Mohorovicic Discontinuity and the basic model of crustal structure. Acta Geophys. Sin., 16 (in Chinese), pp. 1—5

Zhang Bosheng (Chang Bosheng) (1958), The Pre-Sinian sequences and geotectonic development in the Zhongtiaoshan area. Bulletin of Northwest China University, 2, pp. 1—19.

Zhang Bosheng (Chang Bosheng) (1962) The mosaic earth's crust. Acta Geol. Sin., 42(3) (in Chinese), pp. 275—305

Zhang Wenyou (Chang Wenyou) (1960) Stress analysis of the major structural explosive systems of China. Sci. Sin., 19, pp. 604—608

Zhang Wenyou, Ye Hong, Zhong Jiayou (1978) On "fault blocks" and "plates". Sci. Sin., 22(2) (in English), pp. 195—211

Zhang Wenyou, Zhong Jiayou (1977) On the development of fracture systems in China. Sci. Geol. Sin., 3 (in Chinese), pp. 197—209

Zheng Xuezheng, Cong Bolin, Zhang Wenhua, Yan Zheng (1978) Discussion on the petrochemistry of Cenosoic basaltic rocks in Eastern China. Sci. Geol. Sin., 3 (in Chinese), pp. 253—265

Zhou Yaoxiu, Liu Wenjin (1979) Regional gravitational field of China: its basic charateristics. Geophys. Geochem. Explor., 1 (in Chinese), pp. 14—17

Zhu Shixing, Cao Ruiji, Zhao Wenjie, Liang Yuzuo (1978) An outline of the studies on stromatolites from the stratotype section of Sinian suberathem in Chihsien County, North China. Acta Geol. Sin., 52(3) (in Chinese), pp. 209—221

Zhu Xiangyuan, Liu Chun, Ye Sujuan, Lin Jinlu (1977) Remanance of red beds from Linzhou, Xizang and the Northeward Movement of the Indian Plate. Sci. Geol. Sin., 1 (in Chinese), pp. 44—51

Zhu Ying (1979) Geotectonics of the North China Massif and regularity of distribution of Anshan-type iron deposits based on preliminary analysis of aeromagnetic surveys. Geophys. Geochem. Explor., 1 (in Chinese), pp. 3—13

Explanations of Plates

Plate I

Fig. 1. Late Proterozoic Nankou System (Z) unconformably overlying the Early Proterozoic Wutai Group (Pt₁) which in turn overlies the Archean Fuping Granite (the Lanzhishan granitic terrain γ_1)(at Qiyue Village², Wutai County², Shanxi Province, photo by Wu Jiashan 1979)

Fig. 2. Basal conglomerate from the Wutai Group (at Tiebao Village, Wutai County, Shanxi Province, photo courtesy Wu Jiashan 1979)

Fig. 3. Basal part of the Wutai Group (Pt₁¹) unconformably overlying the top part of Archaean Fuping Group (Longquanguan Group, Ar) (at Tiebao Village, Wutai County, Shanxi Province; photo courtesy Wu Jiashan 1979)

Fig. 4. Basal metamorphosed conglomerate from the Wutai Group (at Qiyue Village, Wutai County, Shanxi Province; photo courtesy Wu Jiashan 1979)

Plate II

Fig. 1. Pillow structure of intermediate basic lava from Late Triassic Langjiexue Group in northern Himalayan Eugeosyncline (at Beibaidi Village, Nagarze County, Tibet; photo courtesy Chen Guoming 1978)

Fig. 2. Pillow structure of basic lava from Cambrian-Ordovician sequences in northern Qilian Eugeosyncline (at Yushigou of Qilian Mts., photo courtesy Zhu Zhizhi 1974)

Fig. 3. Pillow structure of Cretaceous basic lava in the Yarlung-Zangbo River Eugeosyncline (in Zhongba County, Tibet: photo courtesy Chen Guoming 1978)

Fig. 4. Spilite from Cambrian-Ordovician sequences in northern Qilian Eugeosyncline (cross Nicol 6.3×3.2) (Provided by Zhu Zhizhi)

Fig. 5. Tholeiite from Cambrian-Ordovician sequences in northern Qilian Eugeosyncline (cross Nicol 6.3×3.2) (Provided by Zhu Zhizhi)

Plate III

Figs.1,2,3. Radilaria from the Himalayan Eugeosynclinal belt (Provided by Chen Guoming)

Fig. 4. Glaucophane-schist from the northern Qilian Eugeosyncline (at Qingshuigou in Qilian Mts.) (Provided by Zhu Zhizhi)

Figs. 5,6. Radilaria from Cambrian-Ordovician siliceous rocks in northern Qilian Eugeosyncline (at Baijingsi in Qilian Mts.) (Provided by Zhu Zhizhi)

Plate IV

Fig. 1. Mélanges distributed along the Yarlung-Zangbo River Suture Line (photo taken with the camera-lens facing southwest; the light-colour swell (positive element) represents exotic block Carbonaceous Limestone (yielding Fusulina), and the gentle low-lying element, late Cretaceous basic lava (in Zhongba County, Tibet; photo courtesy Chen Guoming 1978)

Fig. 2. Mélanges distributed along the Yarlung-Zangbo River Suture Line (with the camera lens facing north, lightcoloured positive element represents exotic block of the Late Triassic Limestone; and the western flank and the low-lying element Jurassic basic lava (at Zhangdong village, Zagarzê County, Tibet; photo courtesy Chen Guoming 1978)

Figs. 3, 4, 6. Flysch imprints from the Triassic Xikang Group in the Bayan Har Miogeosyncline (4 is on the same scale as. 3.) (Chengxi, Xiaojinzi County, Sichuan Province; photo courtesy Ren Jishun 1965)

Fig. 5. Flysch imprints from the Lengjiaxi Group at the basement of the Yangtze Paraplatform (in Taoyuan County, Hunan Province; photo countesy Wang Henian 1961)

Fig. 7. Late Ordovician flysch imprints from W. Zhejiang-S. Anhui Platform-Fold Belt within the Yangtze Paraplatform (in Yuqian County, Zhejiang Province; photo courtesy Ai Changxing 1974).

Plate V

Figs. 1, 3, 4, 5. Late Ordovician flysch imprints from W. Zhejiang-S. Anhui Platform-Fold Belt within the Yangtze Paraplatform (in Yuqian County, Zhejiang Province; photo courtesy Ai Changxing 1974)

Fig. 2. Flysch "disturbance structure" (in Yuqian County, Zhejiang Province; photo courtesy Ai Changxing 1974)

Figs. 6, 7, 9, 10. Flysch imprints from the Lengjiaxi Group (at the basement of the Yangtze Paraplatform) (in Taoyuan County, Hunan Province; photo courtesy Wang Hongnian 1961)

Fig. 8. Flysch "disturbance structure" (in Taoyuan County, Hunan Province; photo courtesy Wang Hongnian 1961)

Plate VI

Figs. 1, 2, 3, 5, 6. Flysch imprints from the Middle and Early Triassic Liufengguan Group in the Qinling Geosyncline (in Fengxian County, Shaanxi Province; photo courtesy Zhu Zhizhi 1963)

Fig. 4. Flysch imprints from the Late Triassic Liangjiexue Group in northen Himalayan Eugeosyncline (in Zetang, Tibet; photo courtesy Chen Guoming 1978)

Fig. 7. Graded bedding in the Middle-Early Triassic flysch from Qinling Geosyncline (Collected by Jiang Chunfa and Zhu Zhizhi 1963)

Fig. 8. Trace fossils from the Middle-Early Triassic flysch from Qinling Geosyncline (Collected by Jiang Chunfa and Zhu Zhizhi 1963)

Plate VII

Figs. 1, 2. Fold structure from the Xikang Group in the foldbelt within Bayan Har Miogeosyncline (Xiaojin County, Sichuan Province; photo courtesy Ren Jishun 1965)

Figs. 3, 4. Triassic folding structure in the Sanjiang Fold System (on the bank of Lancang River in the Dêqên-Weixia area, Yunnan Province; photo courtesy Ren Jishun (Fig. 3.) and Qu Jingchuan (Fig.4.) 1964; it should be noted that the *hammer* shows the mode of occurrence of cleavage, and that the *white line* indicates beddings)

Plate VIII

Fig. 1. Early Triassic folding structure in Qinling Geosyncline (with the camera lens facing west, on a scale of 1:100) (in Fengxian County, Shaanxi Province; photo courtesy Zhu Zhizhi 1963)

Fig. 2. Folding structure composed of the Late Triassic Longjiexue Group in Qinling Eugeosyncline (with the camera lens facing west, on a scale of 1:1000) (in Qiongjie County, Tibet; photo courtesy Chen Guoming 1978)

Fig. 3. Folding structure composed of the Jurassic Nyanyaxungla Group in northern Himalayan Eugeosyncline (with the camera lens facing west, on a scale of 1:300) (in Girong County, Tibet; photo courtesy Chen Guoming)

Fig. 4. Folding structure composed of the Triassic Xikang Group in the Bayan Har Miogeosyncline (at Dawei, Xiaojin County, Sichuan Province; photo courtesy Ren Jishun 1965)

Plate IX

Landsat image mosaic of eastern Asia showing tectonics and geomorphological feature of eastern China (reduced photo after the mosaic given by Dr. Edwar Chao 1977)

Plate X

Landsat image mosaic (on a scale of 1:5,000,000), where the biggest lake is Nam Lake, and the biggest river is the Bulamapute River. The mosaic clearly shows the easterly extension of the Yarlung-Zangbo River Deep Fracture.

Plate XI

Fig. 1. A 1:5,000,000-scale Landsat image mosaic for Altun-Beishan Mt. region on the left lower side of which an ear-shaped lake is known as the Lop Nur. This particular mosaic shows clearly the Altun-Beishan Mt. Fracture Zone, with the Altun Deep Fracture and the Qiemo-River Deep Fracture being most striking.

Fig. 2. A 1:5,000,000-scale Landsat image mosaic for Jiujiang region, the longest river on which is the Changjiang (Yangtze) River with the Ganjiang River and Poyang Lake being located to the south of it, and with Caohu Lake to the north of it. This particular mosaic shows strikingly the southerly extension of the Tancheng-Lujiang Fracture through Changjiang (Yangtze) River and Jiuling Mt.

Plate XII

A 1:2,000,000-scale Landsat image mosaic for Kunming region and for the area south of it, the lake on the northern side of which is Dianchi Lake, still southerly is Fuxian Lake. This particular mosaic shows clearly the Red-River and the Xiaojiang River Deep Fractures overpassing through the Ailao Mt. to join the Dien Bien Phu Fracture (in Vietnam).

Plate XIII

Fig. 1. A 1:5,000,000-scale Landsat photograph mosaic for the Sichuan-Guizhou-Guangxi region, on the northwestern corner of which is the eastern part of Sichuan Basin, and on the southern corner of which is the southern end of the Jiangnan Axis. The present mosaic shows remarkably the salient features of the covering strata of both the Yangtze Paraplatform and the Post-Caledonian South China Paraplatform.

Fig. 2. A 1:5,000,000-scale Landsat photograph mosaic for Zoigê region in northwestern Sichuan, on the western part of Sichuan Basin. The mosaic shows clearly the configuration of the Zoigê Massif and the Longmen Mt. Deep Fracture.

Plate XIV

A 1:2,000,000-scale Landsat photograph mosaic for Youjiang River region, the *upper side* of which indicates approximately the east. The river shown on the mosaic is the Youjiang River. The *black portion* represents the Late Paleozoic carbonate rocks. The mosaic shows clearly a leaner fold composed of the Triassic Pingerguan Group (flysch).

Note: Plates X, XI and XIII are revised versions of Landsat photograph mosaics provided by Dr. Edwar Chao.

Plate I

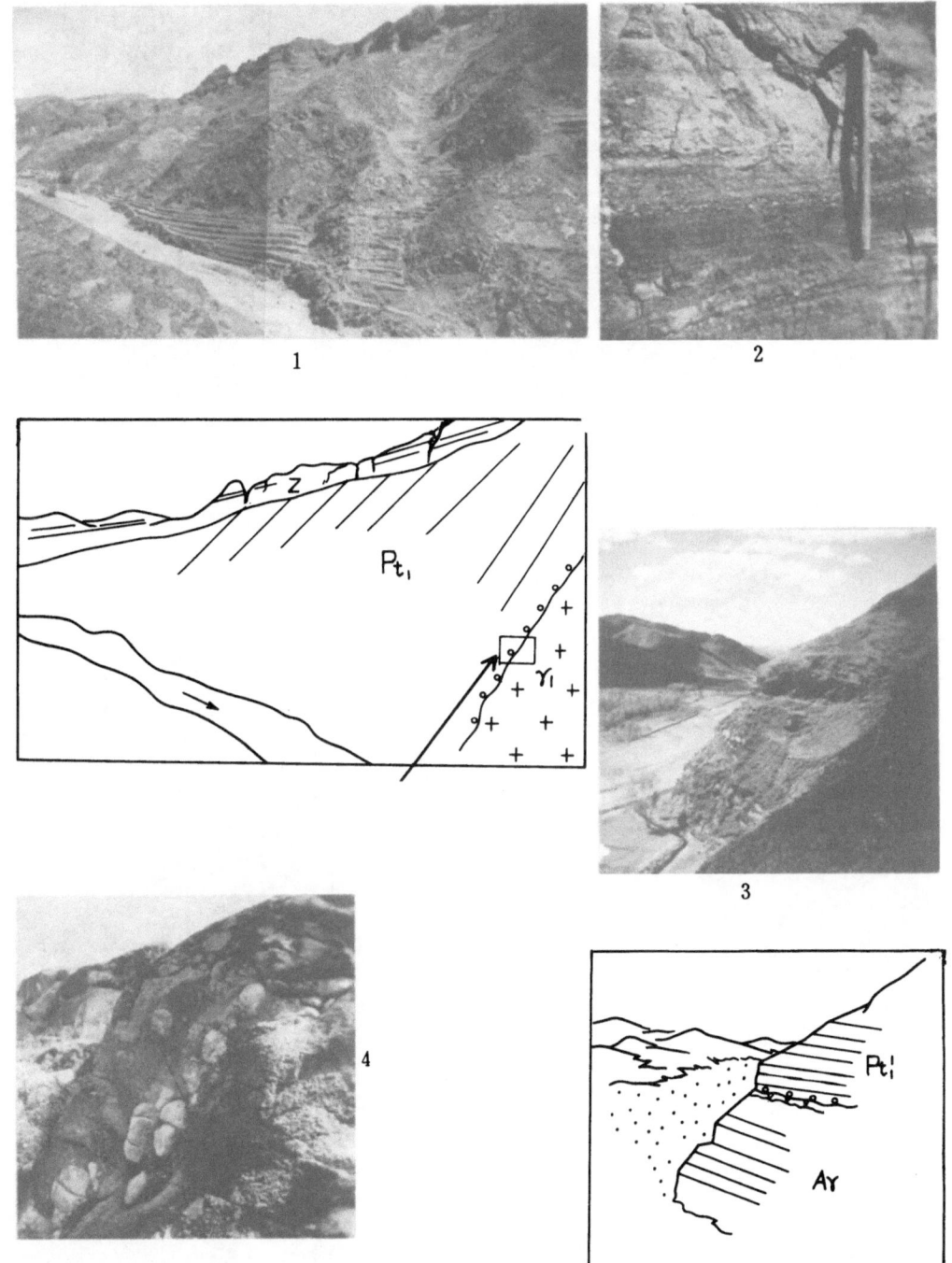

Plate II

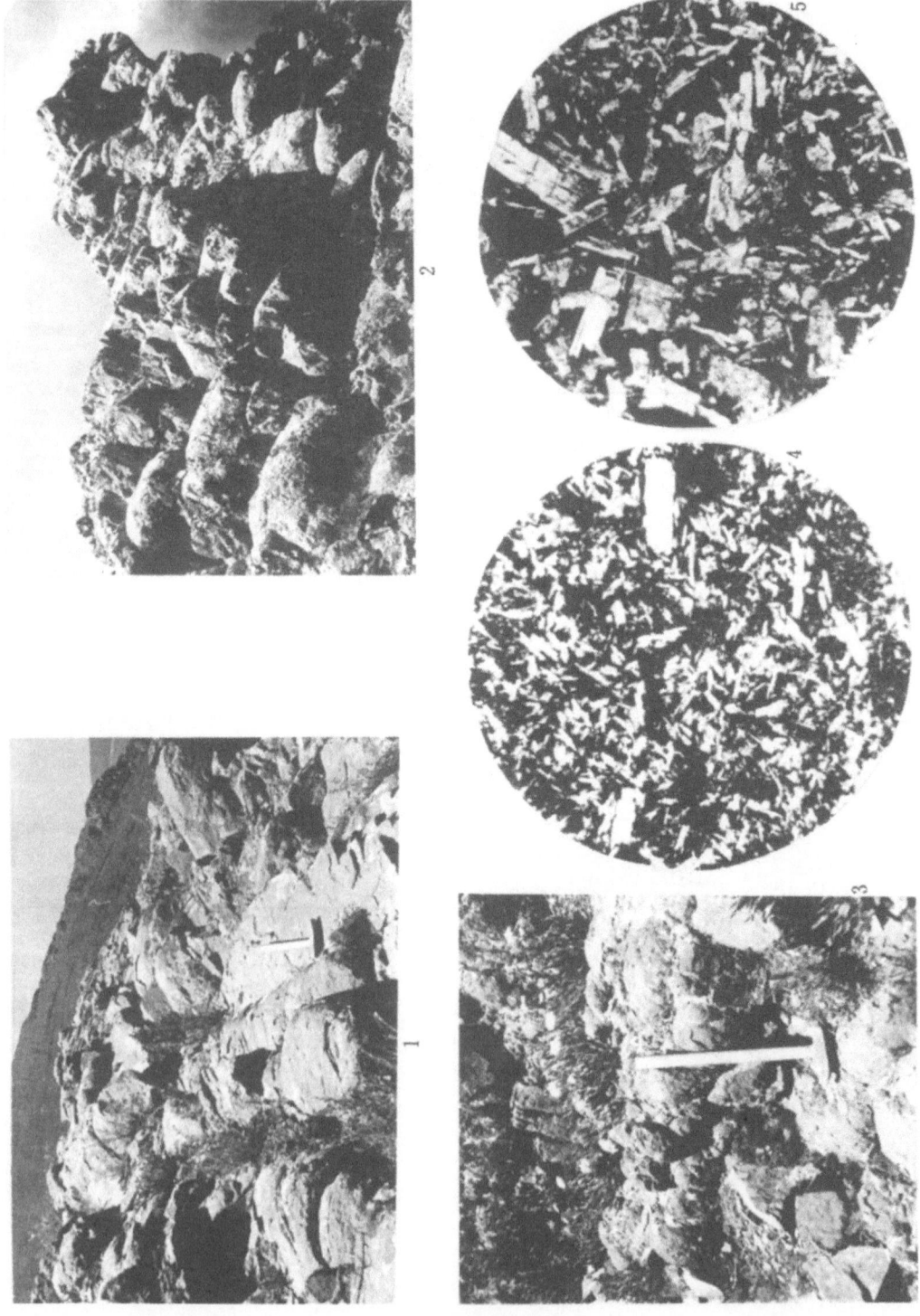

Plate III

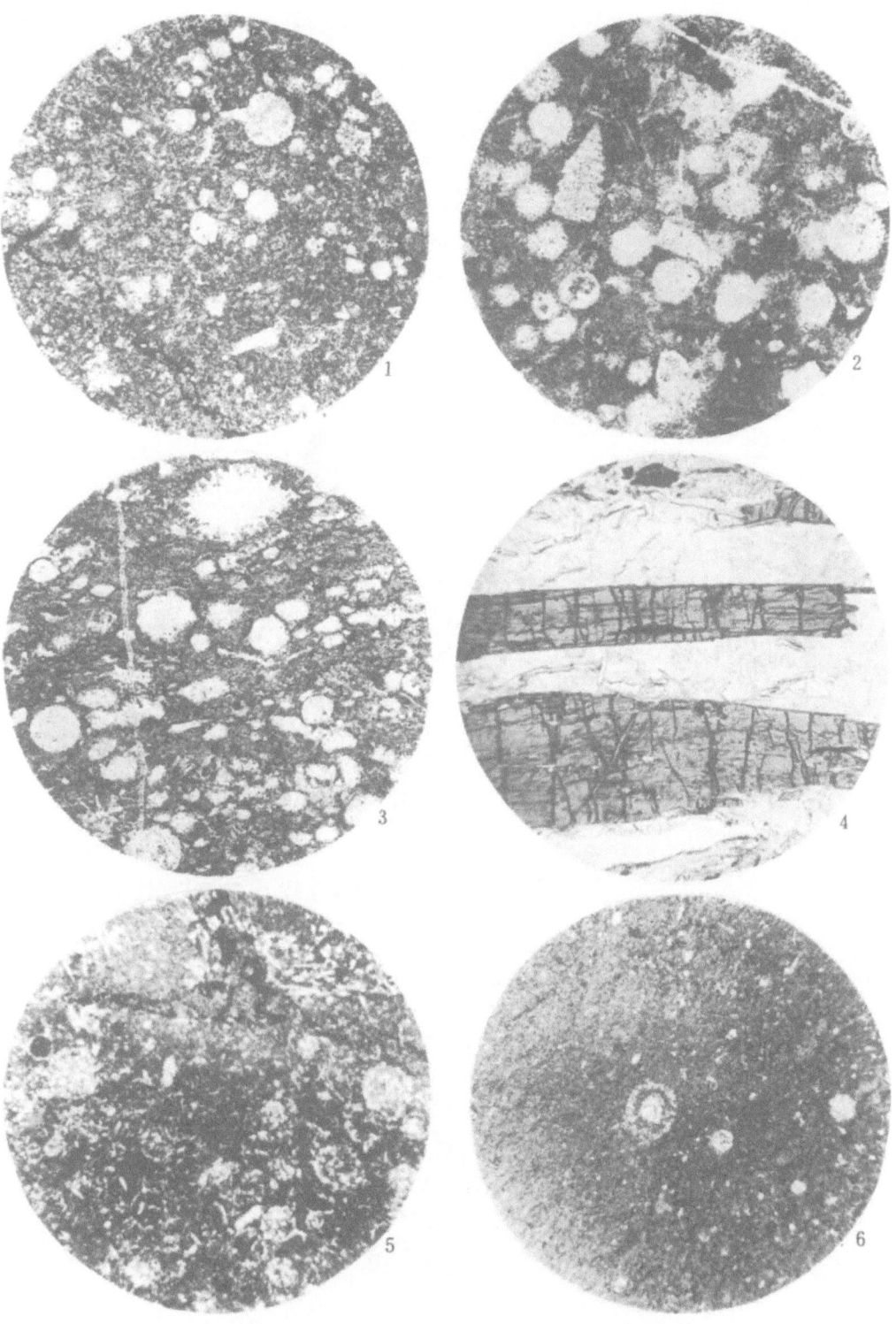

Plate IV

Plate V

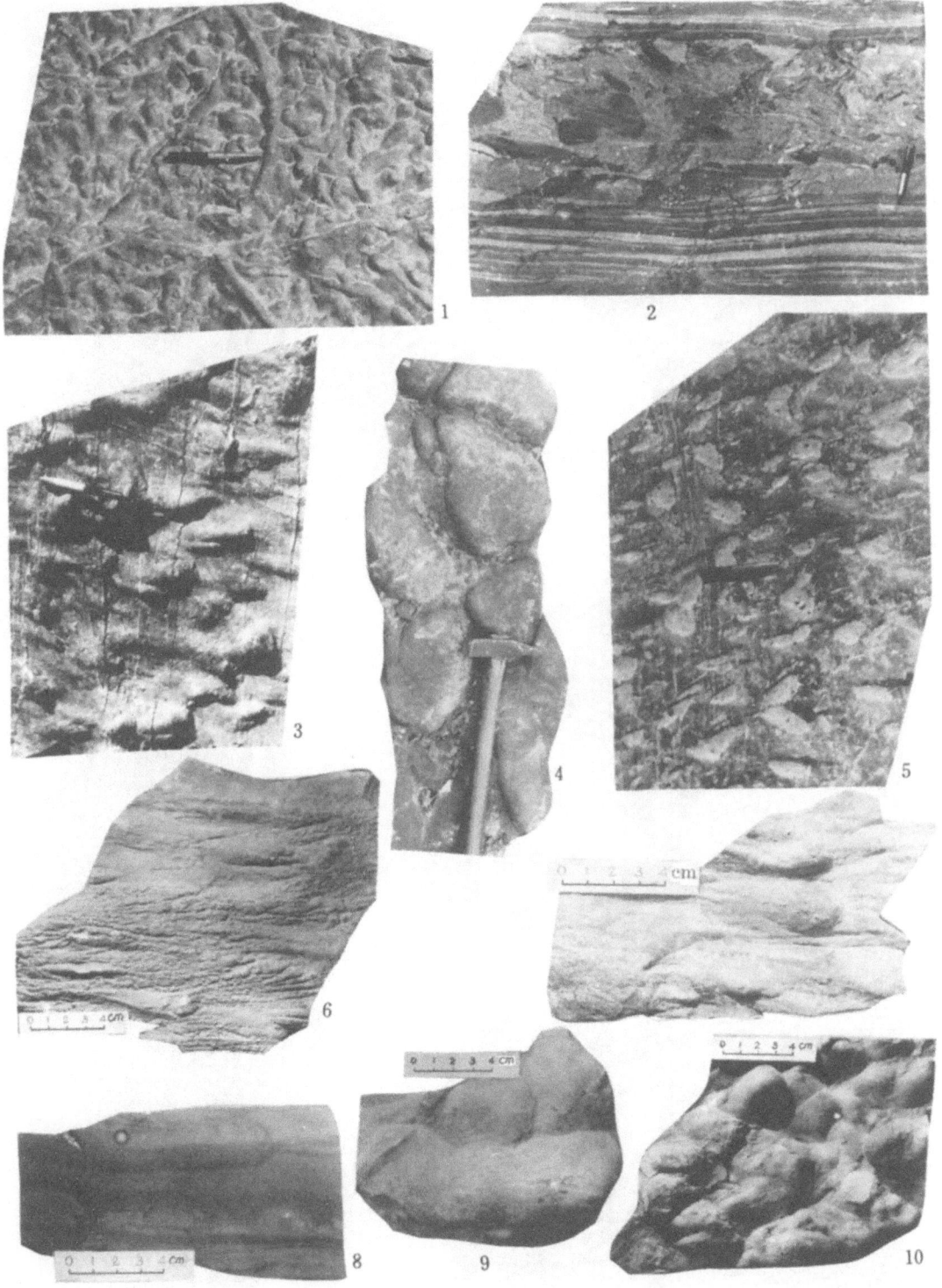

Plate VI

Plate VII

Plate VIII

嘉　　陵　　江

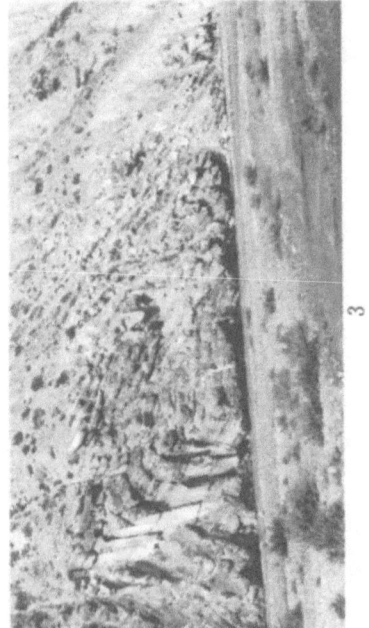

Plate IX

Plate X

Plate XI

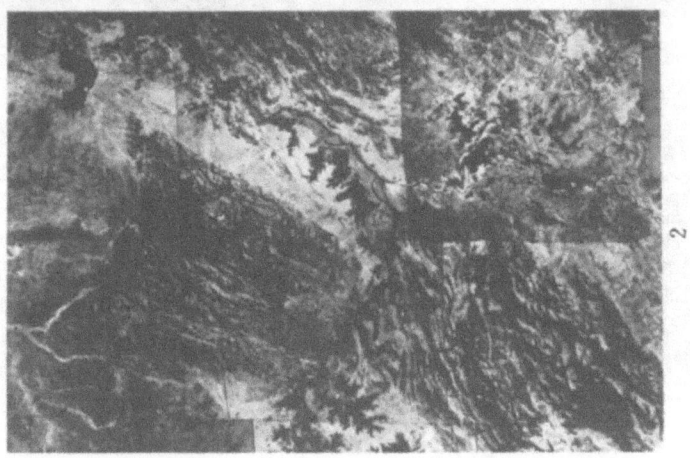

Plate XII

Plate XIII

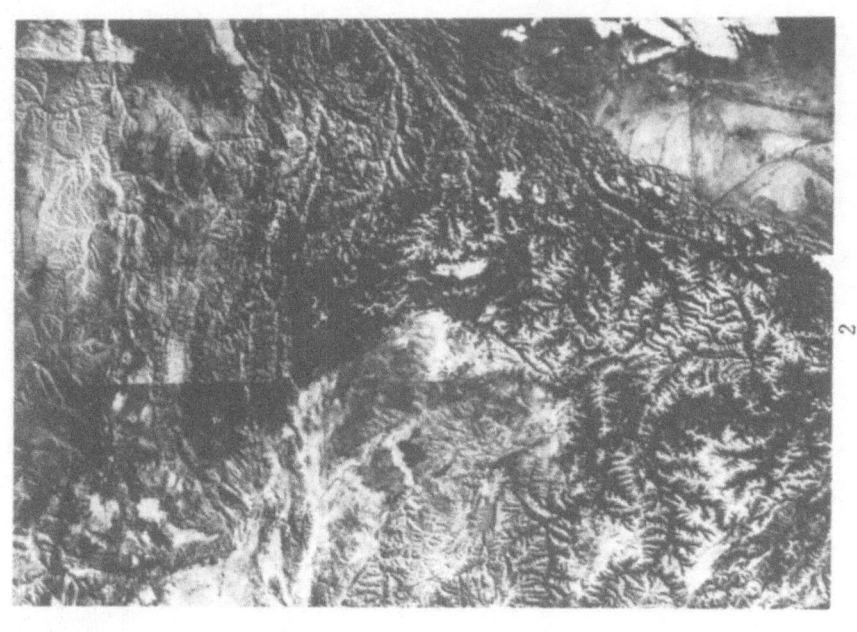

Plate XIV